Contents

Acknowledgements

We would like to thank, in particular, Robert F. Legget, Arden Phair and Wesley B. Turner, who provided helpful comments on the draft manuscript; John Burtniak, who allowed us access to his invaluable collection of postcards; and Sheila Wilson, who prepared the Index.

Many other individuals and institutions have also contributed advice and material: Virginia Anger and staff, Port Colborne Historical and Marine Museum; Loris Gasparotto, cartographer, Department of Geography, Brock University; Carol C. Greene, Town of Sullivan (N.Y.) Historian; Craig and Mary Lou Hanyan; John N. Jackson; Pat McIntyre and staff, National Archives of Canada, Map Division; Divino Mucciante, photographer, Technical Services, Brock University; Arden Phair, curator, and staff, St. Catharines Historical Museum; the late Francis Petrie; Alfred F. Sagon-King; Robert Shipley, Welland Canals Society; G.D. Shutlak, Map Archivist, Public Archives of Nova Scotia; the staff of Special Collections at the St. Catharines Centennial Library; R.M. Swackhammer and staff, Welland Historical Museum; Alan Sykes; John Ter Horst and other staff in Maintenance Engineering, St. Lawrence Seaway Authority, Western Region; Sharon Uno and staff, National Archives of Canada, Photography Division. The staffs of the Cayuga, Dunnville, Thorold and Welland public libraries, the Metropolitan Toronto Reference Library's Baldwin Room, and the Ontario Archives, have also been unfailingly kind and helpful.

We are grateful to the Social Sciences and Humanities Research Council, and to the Ontario Arts Council, for their generous assistance in support of the photographic research necessary for this book.

Abbreviations:	
AO	Archives of Ontario
NAC	National Archives of Canada (formerly Public Archives of Canada)
NMC	National Map Collection
PANS	Archives of Nova Scotia
PCHMM	Port Colborne Historical and Marine Museum
ROM	Royal Ontario Museum
SCHM	St. Catharines Historical Museum
WHM	Welland Historical Museum

MR. MERRITT'S DITCH

A Welland Canals Album

The Four Welland Canals at Thorold, Ontario: Mr. Merritt's Ditch was a response to the problem of getting ships over the Niagara Escarpment. The remarkable series of solutions to this problem — the four Welland Canals — have transformed the landscape with a network of linked waterways. This aerial view shows the Fourth Canal cutting across the centre, with a laker in Lock 6 heading up the twinned flight locks to Lake Erie. Lock 7 is on the left. One lock of the Third Canal can be seen in the large pond, and others appear as the canal winds away at bottom right. The gently curving bare patches at the top of the photograph mark the route of the First and Second Canals. The First Canal altered the line of settlement in the peninsula, with profound and abiding consequences for the economy of the area. —St. Lawrence Seaway Authority: Western Region

MR. MERRITT'S DITCH

A Welland Canals Album

Roberta M. Styran and Robert R. Taylor

Stoddart

A BOSTON MILLS PRESS BOOK

Canadian Cataloguing in Publication Data

Styran, Roberta McAfee, 1927-
  Mr. Merritt's ditch: a Welland Canals album

Includes bibliographical references and index.
ISBN 1-55046-005-6

1. Welland Canal (Ont.) – History. I. Taylor,
Robert R., 1939-   . II. Title.

HE401.W4S89 1992   386'.47'0971338   C92-093920-1

Edited by Noel Hudson
Cover design by Gillian Stead
Typography by Lexigraf
Printed in Canada

First published in 1992 by:
Stoddart Publishing Co. Ltd.
34 Lesmill Road
Toronto, Ontario Canada M3B 2T6
(416) 445-3333 Fax: (416) 445-5967

A BOSTON MILLS PRESS BOOK
132 Main Street
Erin, Ontario N0B 1T0
(519) 833-2407 Fax: (519) 833-2195

Winners of the
Heritage Canada
Communications Award

American Association
for State and Local History
Award Winner

The publisher wishes to acknowledge the support of the Canada
Council, the Ontario Arts Council and the Ontario Publishing
Centre in the development of writing and publishing in Canada.

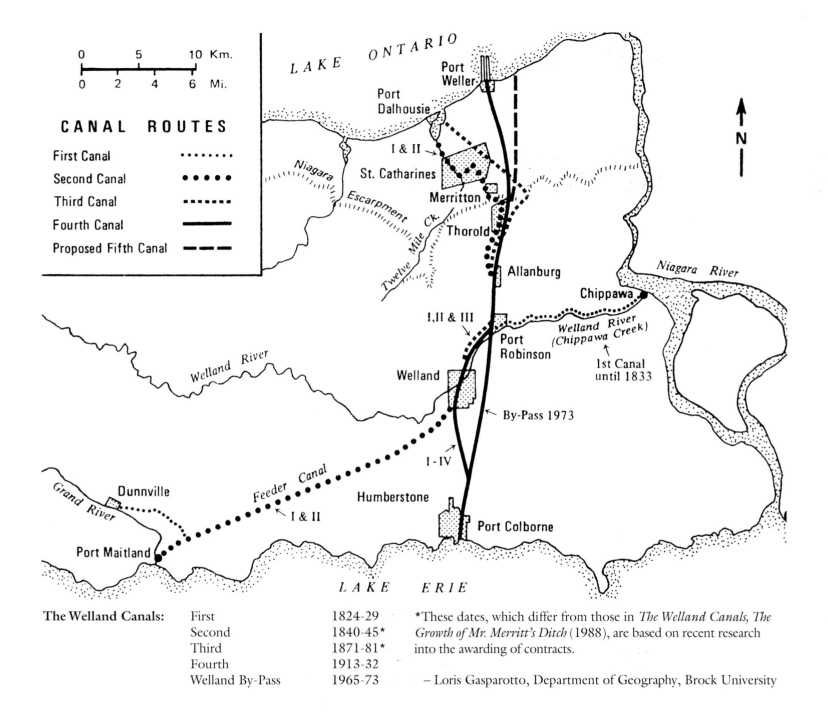

CANAL ROUTES

First Canal · · · · · · · ·
Second Canal ● ● ● ● ●
Third Canal ▪ ▪ ▪ ▪ ▪
Fourth Canal ▬▬▬▬▬
Proposed Fifth Canal ▬ ▬ ▬

The Welland Canals:

First	1824-29	
Second	1840-45*	
Third	1871-81*	
Fourth	1913-32	
Welland By-Pass	1965-73	

*These dates, which differ from those in *The Welland Canals, The Growth of Mr. Merritt's Ditch* (1988), are based on recent research into the awarding of contracts.

– Loris Gasparotto, Department of Geography, Brock University

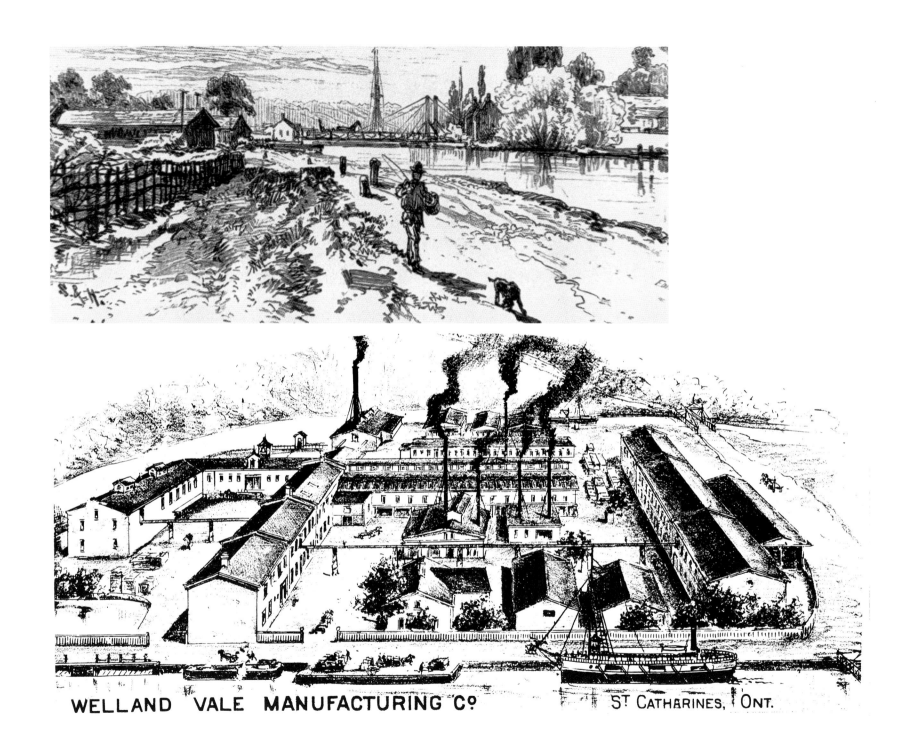

WELLAND VALE MANUFACTURING C.º   S.T CATHARINES, ONT.

INTRODUCTION

"The Welland Canal (1824-29) was initiated by local businessmen to stimulate local and regional trade. By the 1840s the canal's importance to the economy of British North America was recognized by government takeover of the privately owned Welland Canal Company in order to finance the urgently needed rebuilding. The locks and channel were enlarged (Second Canal, 1842-45) to accommodate the increasing size of ships. Following Confederation (1867) and the opening of the Canadian West, the waterway was seen as a vital link in a crucial artery of North American trade, and was once again rebuilt and enlarged (Third Canal, 1874-87). By 1913 the ever-larger ships from around the world necessitated yet another reconstruction (Fourth Canal, 1913-32). These are the bare facts. The history of the Welland Canal is actually a saga of men, money, machines, misery, and daring flights of technological imagination."

So we wrote in *The Welland Canals: The Growth of Mr. Merritt's Ditch* (1988) [p 13]. In the course of our research for that volume we located far more material than could be included. Moreover, some themes introduced in the earlier book cried out for further development and new themes had come to mind. Thus was born the idea of this companion volume.

One continuing theme which now has been given a major focus is the contrast between romance and reality. In our first book we tried to picture both and pointed out some of the pitfalls of accepting the "romantic" view without careful thought. Since the romantic attitude to landscape (in which the beauty and grandeur of nature overwhelmed puny mankind) was a feature of late 18th and early 19th century art in both Europe and North America, it is scarcely surprising that those who portrayed the early Welland Canal saw it in that light [Int. 1]. Because artists working with pencil, oils or watercolours sometimes produced distorted (or contradictory) views of the canals, we occasionally found it difficult to decide just how the waterway and its surroundings actually looked [Int. 2].

Int. 1: Welland Vale, St. Catharines, 1882 (opposite top): This charming view of an area near Lock 2 of the Second Canal is typical of the late 19th-century "romantic" view of the world. Artists' renderings often presented picturesque, bucolic scenes wherein the canal seemed to meander like a lazy stream in a Homer Watson painting, more a gently flowing river than a man-made channel vitally important to the economies of Canada and the United States. – *Picturesque Canada*

Int. 2: Welland Vale, St. Catharines, ca. 1890 (opposite bottom): The artist, employed to portray various civic, industrial and commercial establishments for a "bird's eye" map of St. Catharines, presents a more realistic view of this important Second Canal area than that shown in Fig. Int.1. Here is a bustling, smoky — and presumably noisy — scene. Note that the bridge in the lower right corner here appears in the centre background of Int. 1. Even here, however, we cannot be sure that this is an accurate depiction. – St. Catharines Centennial Library

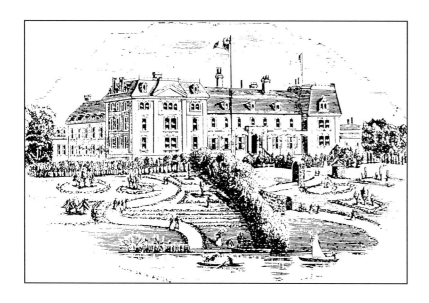

Int. 3 and 4 St. Catharines, ca. 1870 and 1865: While the architectural details of the Springbank Hotel are well rendered in the drawing, the topography is not. The artist's impression of a gently sloping and romantically landscaped bank down to the canalized Twelve Mile Creek is belied by the photograph, which gives a much more realistic sense of the steepness of the terrain. – *St. Catharines Centennial*, St. Catharines 1876; SCHM

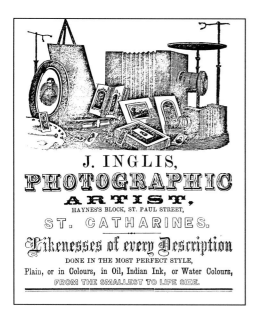

Int. 5 St. Catharines, 1865: Mr. Inglis was obviously proud of his ability to portray both people and scenery in an artistic fashion. Nevertheless, he and his colleagues, then and since, have left an invaluable photographic record of so many long-vanished scenes. – Mitchell's *General Directory of the Town of St. Catharines, 1865*

Int. 6 St. Catharines, 24 July, 1915 (opposite): The site of the future Lock 2 and what was then an up-to-date mechanical shovel used in its excavation were the principle interests of the photographer. The men, with their carts and teams of animals employed to haul away the spoil, were also an important component of canal building before the 1920s, when the use of steam- and gasoline-powered vehicles became more common. But the men and the hot and dusty conditions in which they laboured were irrelevant considerations to both the official photographers and their employers. The "official record," then, can tell us little about working conditions, and about accidents, especially those involving injuries or death. True, the activities of some of the riggers, pitmen, drill operators, and other labourers, who literally risked their lives building the present waterway, were captured on film, but this was essentially an incidental record. – SCHM: N-1001

By the late 19th century widespread use of photography provided the means to achieve a more accurate record of the canals and canal-side scenes [Int. 3, 4 and 5] (another theme elaborated upon in this volume). Thousands of official photographs of Fourth Canal construction provide a detailed and technically excellent record, still consulted by St. Lawrence Seaway engineers [Int. 6]. But while engineers can rejoice in this wealth of material, the social historian bemoans the lack of specific information on working conditions — in part a result of the photographers' choice of subject.

As we grappled with the interpretative problems inherent in the *visual* representations of the canals, we were forced to confront similar difficulties with *written* accounts, such as the biography of William Hamilton Merritt, written by his son Jedediah [Int. 7]. Even the Welland Canal Company Papers, the Register of Deeds [Int. 8], and the Merritt Papers (the primary sources) provide only an "official" record, often failing to answer the historians' questions about the people involved and the conditions under which they worked.

Int. 7 Jedediah Prendergast Merritt (1820-1900), ca. 1876: The second son of William Hamilton Merritt wrote a detailed biography of his father, published in 1875. It included copious quotations from his father's journal and family correspondence, as well as from newspapers of the time, and is an essential starting point for any understanding of W.H. Merritt and his manifold business and political interests. Nevertheless, as a personal account compiled by an admiring son, it is far from being a complete record of the man or of his life and times. For example, in 1836 the government of Upper Canada was persuaded by William Lyon Mackenzie to investigate his charges of malfeasance against Merritt and other officers of the Welland Canal Company. The biography's main reference to the final *Report* was a letter from Merritt to his wife, in which he said: "The farce has ended. And after being tried by our enemies, we have been acquitted with credit" [*Biography*, p. 159]. The *Report* itself concluded that "for some years past the affairs of the Company have been conducted in a very loose and unsatisfactory manner, which may have, and no doubt has, originated in their being frequently much cramped for means to carry on the necessary repairs . . . ; and when your committee take into consideration the magnitude of the undertaking, and the many unforseen disadvantages the Direc-tors have had to struggle against, they feel inclined to put the most favorable construction upon their general conduct, and to acquit them of any *intentional* abuse of the powers vested in them although it is difficult for your committee to account for or excuse their conduct in the sale of the Hydraulic Works . . ."   – [*Report*, p. 9]. A difference of emphasis, perhaps. SCHM: N-2483

Int. 8 Register of Deeds B, 1851: This clause, excerpted from the deed between the Commissioners of Public Works and Robert Lawrie *et al.*, regarding their grist mill at Port Dalhousie, indicates some of the conditions imposed upon mill-owners. The vagueness of such documents adds to the frustration of researchers wanting to know about the "driving wheels" and other equipment of early mills.   – Board of Public Works, *Register B, 1841-1880*

Of particular frustration to the historian, industrial archaeologist, and canal buff is the lack of any visual record of the original Welland Canal, other than the surveys of lands appropriated. There is not even much in the way of detailed written descriptions. Both the Rideau and Erie Canals, for example, were government undertakings (British and New York State respectively) and involved construction over several hundred miles. The men responsible for those canals required both visual and written accounts of the work. In contrast, progress on the privately owned Welland was easily monitored by any member of the Welland Canal Company who chose to ride along its relatively short route. Hence the best we can do today is to deduce, from representations of other canals of the period, what conditions might have been on the Welland [Int. 9, 10 and 11].

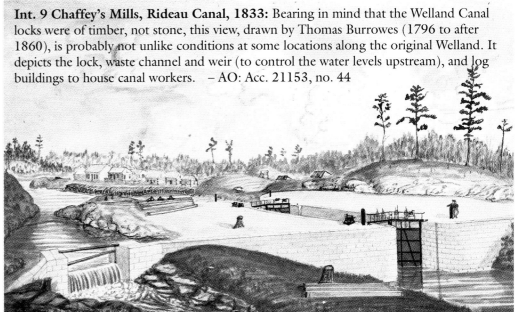

Int. 9 Chaffey's Mills, Rideau Canal, 1833: Bearing in mind that the Welland Canal locks were of timber, not stone, this view, drawn by Thomas Burrowes (1796 to after 1860), is probably not unlike conditions at some locations along the original Welland. It depicts the lock, waste channel and weir (to control the water levels upstream), and log buildings to house canal workers. – AO: Acc. 21153, no. 44

Int. 10 West Troy, N.Y., 1830-32: This romantic and seemingly peaceful scene at the locks where the Erie and Champlain Canals met was painted by John Hill. While the barges depicted were rarely, if ever, seen on the Welland, the view gives us some sense of First Canal's surroundings and atmosphere (though certainly not of the sounds and stench of a typical canal scene of the period). – Courtesy of the New-York Historical Society, New York City

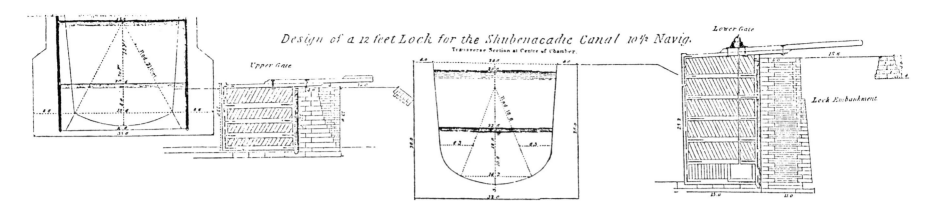

Int. 11 Lock Design for the Shubenacadie Canal, N.S., ca. 1828: Scottish-born Francis Hall, trained by the famous British civil engineer Thomas Telford, worked at one time or another on canal surveys and construction in four of the British North American colonies. While Hall was consulted by the Welland Canal Company on more than one occasion, no technical drawings for the Welland have survived. We can only envy researchers into the history of the Shubenacadie, since many surveys and drawings which Hall produced during its construction have been preserved. – PANS: MG 24, Vol. 1

Other themes from the first volume continue to be featured, such as the international context of the Welland [Int. 11 and 12]: British and American engineers produced surveys and plans for the first two canals; the contractors of the First Canal were predominantly American, and the canal labourers in more recent times have haddiverse ethnic backgrounds. The waterway has been, and continues to be, of international importance, both politically and economically. The role of the canal in the development of the Niagara Peninsula [Int. 13 and 14] and the human element in its many facets economic, social, recreational, and so on, are also given prominence.

Int. 12 Nathan S. Roberts (1776-1852): Roberts was just one of many Americans who brought skills and knowledge acquired on the Erie Canal (1817-1825) to assist in construction of the first Welland Canal. Originally a schoolteacher, Roberts taught himself surveying after being persuaded to join chief engineer Benjamin Wright as assistant in 1816. One of his tasks was to get the Erie up the rock face at Lockport, N.Y. This was accomplished by an imposing double set of five stone locks (the stonework of one set may still be admired). When Wright (originally a judge, who became pre-eminent among the self-taught Erie engineers) was approached by William Hamilton Merritt regarding Roberts, he replied, in a letter of 1 October 1824: "I say with pleasure that he is a prudent careful man and free from any visionary plans of internal improvement. Mr. Roberts commenced with me as assistant on the Erie Canal in 1816, and has continued in the employ of the State of New York ever since. . . . I would place every reliance upon his accuracy and care, and . . . can freely recommend him as worthy of every confidence in his profession." – [Printed in the *Report of a Hearing before a Select Committee of the Upper Canada Assembly*, 7 December 1825.]

Roberts was hired by the Welland Canal Company to report on the survey and estimates submitted by Samuel Clowes, and on Francis Hall's estimates for the Deep Cut, in early 1824, and submitted a lengthy report in which he provided detailed reasons for favouring one of the proposed routes over the other [see Chap. 2, Fig. 2.6 for discussion of alternate routes]. – Erie Canal Museum, Syracuse, N.Y.

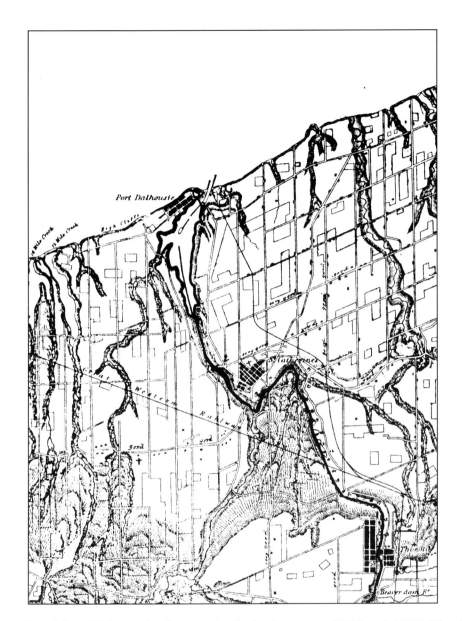

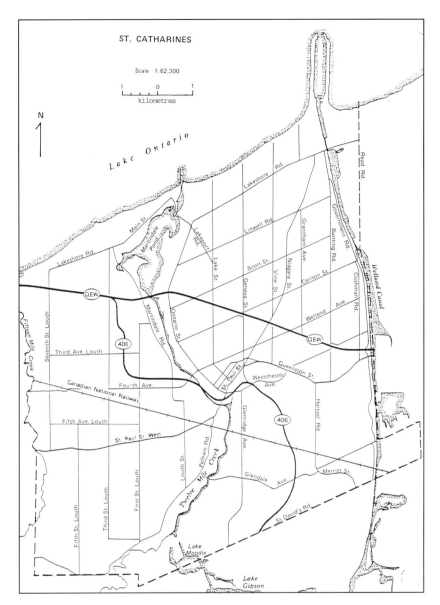

Int. 13 and 14 Port Dalhousie-St. Catharines area, 1866 and 1988: The detail from an 1866 map produced by the British War Office shows both the area topography (especially the Niagara Escarpment and the many streams which flowed north over it to Lake Ontario) and the small but growing communities which owed their being to the First and Second Canals. The impetus given by this initial development was more than sufficient to allow continuing growth in St. Catharines, even when the Third (and later, the Fourth) Canal was routed well away from its centre. Both Port Dalhousie and Merritton were amalgamated with St. Catharines in 1961, by which time Merritton was long past its heyday as an industrial corridor.
– War Office, Southampton; Map Library, Brock University

While we did previously discuss the canals' effect on the changing landscape of the Peninsula, we now include material on how that effect can continue long after the earlier canals have ceased to function as such [Int. 15 and 16]. Another theme barely touched upon previously, and now given greater emphasis, is the role of the Feeder Canal and its communities [Int. 15 and 16].

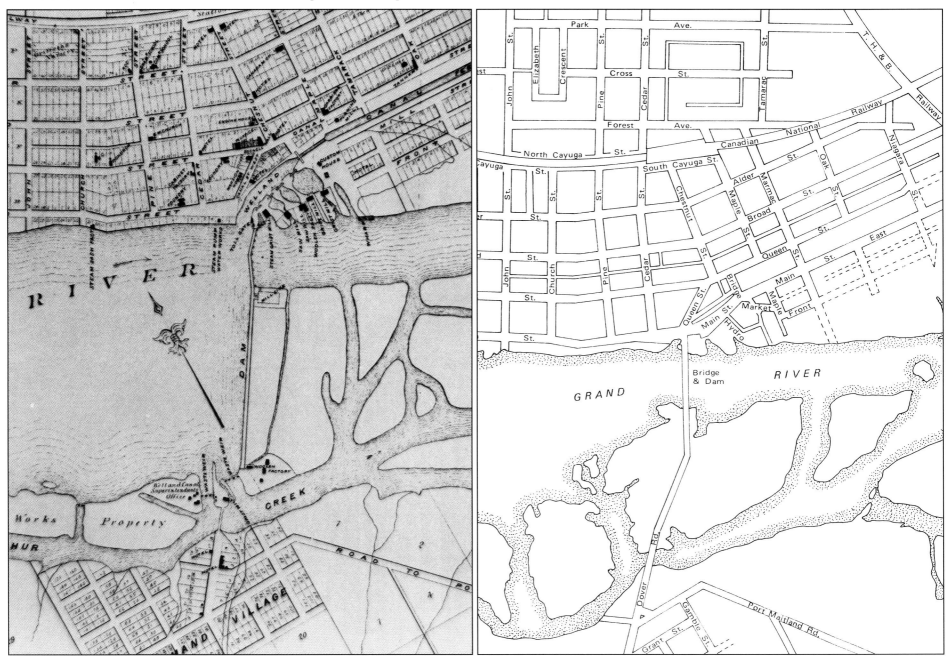

A new theme was suggested by both technical drawings (especially for the Third and Fourth canals) and the official Welland Ship Canal photographs: the remarkable, often surprising, beauty of structures intended, not for aesthetic pleasure, but for the practical fulfillment of a technical function [Int. 17 and 18]. Simply put, many of the bridges, aqueducts, control buildings, and weirs of the canals are often beautiful, both in the technical drawings and in structural form.

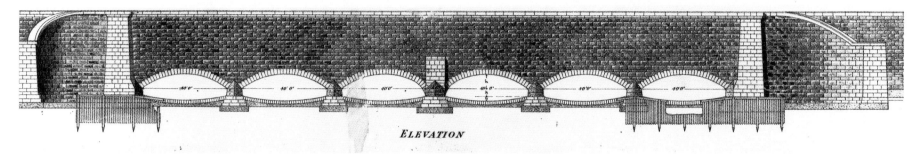

ELEVATION

Int. 17 Elevation for Third Canal Aqueduct, 1886: A typical drawing of a Third Canal structure, intended for the eyes of engineers and contractors only, in itself gives pleasure to the eye, while the stonework of the completed aqueduct is equally satisfying in an aesthetic sense. And yet the purpose of both was purely practical and functional. Both the drawings and the stone structure's remains deserve to be better known. – NAC: NMC-21825

Int. 15 and 16 Dunnville, 1879 and 1986 (opposite)**:** These two maps illustrate one of the themes discussed in Chapter 11, the effect of the successive canals on the landscape of the Niagara Peninsula, even when physical traces of the canal itself may have disappeared. In Dunnville, not only have streets been renamed or disappeared, but the course of the Feeder Canal has become Main Street. Dunnville today shows little evidence of the busy industrial centre it was in the second half of the 19th century — thanks to the Feeder Canal, which supplied the main canal with water from the Grand River from 1828 to 1887.
– Page's *Historical Atlas for the Counties of Haldimand and Norfolk*; Loris Gasparotto, Department of Geography, Brock University

Int. 18 Lock 1, Port Weller, 1925: For the Fourth Canal, the hand-cut stonework of the Second and Third canals gave way to reinforced concrete, a material not generally regarded as a thing of beauty. But who can deny the intrinsic beauty captured here, much of which is never seen by any but engineers, contractors and workmen — and photographers such as J.A. McDonald. The simplicity of the massive lock walls, the perspective towards the far gates, and the precise symmetry of curves and angles of the mitre sill and dwarf walls (for guiding cables of the gate-operating gear) appeal to our visual sense as they did to McDonald's. He preserved a factual record of construction in progress at a particular site, but at the same time revealed and produced a work of art.

– SCHM: N-6823

18

Another new theme which we felt deserved a chapter to itself was that of advertising and promotion [Int. 19]. The Welland Canal Company and (after 1841) the provincial government, then the Government of Canada, industries and businesses of various kinds, and the canal-side communities, all promoted their interests by means of newspapers, directories and gazetters, maps and postcards, even postage stamps.

We have also introduced a section on "Losses" – featuring some aspects of our canal-related heritage that disappeared before, or in spite of, the preservation movement of recent years [Int. 20].

Although we have retained some of the same chapter titles, the present volume differs from the previous one in a number of ways. The first and perhaps most obvious change is the format. We invite our readers to imagine that this is a scrapbook or album assembled over the 40-year career of a Welland Canal engineer with a strong interest in the early canals, a man who worked on the building of the present Ship Canal, experienced the development of the St. Lawrence Seaway, and saw the construction of the Welland By-Pass. The scrapbook, we imagine, was found in a box in his attic after his death. This scenario is not as fanciful as it may sound, since on three occasions in our research we have been fortunate to use old photograph albums, one discovered literally in an attic, another actually put together by a Fourth Canal construction engineer, and the third rescued just as it was about to be dumped in the garbage.

The second change is the fact that, aside from the Introduction, there is very little text, hence the captions are generally quite lengthy. Thirdly, Chapter 12 reflects the heightened interest in the preservation and utilization of canal remnants and canal-related structures. Also, a chapter has been included describing pre-canal conditions in the Niagara Peninsula, which serves to emphasize the Welland Canals' tremendous impact. Finally, a Glossary of canal and canal-related terms has been appended.

We have been greatly encouraged by the response to our first publication (including the corrections) and hope that our readers will derive as much pleasure from this volume. Happy gongoozing! [See Glossary]

R. G. BLAMEY,
SAIL MAKER AND RIGGER,
AND
Importer of Sail-cloths, Cordage, Twines, Bunting, &c.
CANAL,
NEAR THE BUFFALO AND GODERICH RAILWAY,
PORT COLBORNE, C. W.

☞ Ship and Steamboat Colors, Surveyors' and Shooting Tents, made to order, and supplied on the shortest notice.

Int. 19 Port Colborne, 1865 (top right): Shipbuilding flourished along both the main Welland Canal and the Feeder Canal throughout the 19th century [see Chap. 3, Figs. 7-9], and sail-making establishments became a vital subsidiary industry. R.G. Blamey is typical of the hundreds of advertisers who trumpeted their wares in newspapers, directories and gazetteers, proudly proclaiming their locations on or near the canal. – Mitchell's *General Directory for the Town of St. Catharines . . .*, 1865

Int. 20 St. Catharines, 1975: While these small houses (built in the 1850s by Louis Shickluna for men employed in his shipyard) were never intended to be particularly attractive, they too were not unpleasing in their proportions. Though not of great architectural importance, they were among the few remaining structures of the period which survived to the 1970s, and were a vital link with the early Welland Canals, to which St. Catharines owes so much of its prosperity. But they, like so many other canal-related structures, have been swept away, and with them, a part of our heritage. – SCHM: N-1559

Chapter 1: Niagara Before the Welland Canal

1.1 Niagara Peninsula Settlement Patterns, 1825 and 1850: Because water was the chief means of transportation, the early settlements were strung along the Niagara River, from Newark (now Niagara-on-the-Lake) on Lake Ontario, through Queenston, on through what is now Niagara Falls, and then Chippawa to Fort Erie; or they were located where an east-west Indian trail crossed streams running down over the Niagara Escarpment to Lake Ontario. The Short Hills area and other locations where a fall of water might power a mill were also settled. By 1850 the picture had changed: some of the early communities were already showing signs of decline, and the new line of growing settlements was along the course of the Welland Canal (completed 1829, enlarged in the 1840s). – Loris Gasparotto, Department of Geography, Brock University

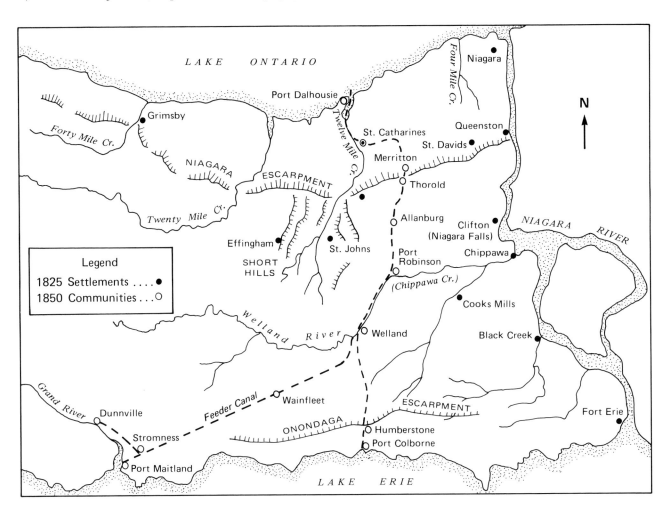

1.2 Queenston, 1790s: Queenston, nestled at the foot of the Niagara Escarpment where it meets the Niagara River, was an important staging place in the "New Portage" which the British had built on their side of the river in 1788 [see Fig. 1.7]. Tensions between Britain and the United States made defence of the new Loyalist setttlements a major concern. Hence the appointment of one of the most capable regimental commanders of the American Revolutionary War, Col. John Graves Simcoe (1752-1806), as the colony's first Lieutenant-Governor, and the military encampment high on the cliff above the village, as shown here in a sketch by the Governor's wife, Elizabeth Simcoe (1766-1850). – AO: Simcoe 66

Mrs. Simcoe kept a diary of her experiences in the new colony, which she illustrated with charming sketches in the romantic manner of the period. Nevertheless, her emphasis on the puniness of man and his fledgling communities in the vastness of nature reflects a basic fact of pioneer life in the Niagara Peninsula: the pattern of settlement was dictated by water transportation in general, and especially by the Niagara River [see Fig. 1.1].

A decade after Mrs. Simcoe's sketch, George Heriot [see Fig. 1.8] described it as follows: "Queenston is a neat and flourishing place, distinguished by the beauty and grandeur of its situation. Here all the merchandise and stores for the upper part of the province are landed from the vessels in which they have been conveyed from Kingston, and transported in waggons to Chippawa, a distance of ten miles, the falls, and the rapids and broken course of the river, rendering the navigation impossible for that space. Between Niagara and Queenston the river affords, in every part, a noble harbour for vessels, the water being deep, the stream not too powerful, the anchorage good, and the banks on either side of considerable altitude" [*Travels Through the Canadas* . . . . London, 1807, p. 157].

1.3 Col. John Graves Simcoe and Mrs. Elizabeth Simcoe: Commander of the Queen's Rangers during the American Revolution, Col. Simcoe was appointed as Lieutenant-Governor in 1791, arriving with his wife the following year. A man of practical vision, he saw the Niagara Peninsula as the future centre of both the colony and its trade with the continent's interior. He also saw the need for immigration to stimulate economic growth, for effective defence of the nascent settlements, and for the opening of roads to facilitate both trade and defence [see Fig. 1.4].

As Governor, Simcoe travelled extensively, often walking for most of his journey: Mrs. Simcoe noted that a return trip to Detroit had taken five weeks. Throughout her diary she described both the pleasures and vicissitudes of travel by land and water.

It is certainly necessary to have a horse that knows the country to cross the bridges we met with everywhere. Some were across creeks and some across swamps. The bridges were made of the trunks of trees of unequal size. The logs are laid loosely across pieces of timber placed length ways. Rotten logs sometimes give way. The horses leg can slip through and is in danger of being broken. The horse I am riding now once had a fall through an old bridge. He goes very carefully now.
– Diary, p. 104; Metropolitan Toronto Reference Library: JRR-3370; NAC: C-11227

1.4 The Corduroy Road Between York and Burlington, 1830: One of Gov. Simcoe's roads was sketched by James Patterson Cockburn (ca. 1778-1847), a soldier and topographical artist who travelled widely in the British North American colonies. This particular road connected the new capital of Upper Canada, York (now Toronto), with the head of navigation on Lake Ontario. Cockburn's watercolour probably gives us a fairly accurate view of a corduroy road in *good* weather, since he and other military artists [see Figs. 1.5, 1.6 and 1.11] were trained to observe and record details of the areas in which they served. Such drawings were essential for planning military strategy in the pre-photography era. This drawing is inscribed on the back: "Hemlock swamps between Yorke and Burlington 1830," which reminds us of the ever-present summer agues and fevers until such swamps were drained. Imagine bumping over the succession of logs laid side by side in any sort of wheeled vehicle without springs, or trying to keep one's horse from accident!
– ROM: 949.39.9

1.5 On a Bush Farm near Chatham, Upper Canada, 1838: Philip John Bainbrigge (1817-1881), another British military artist who depicted colonial pioneer life, left this sketch. It typifies the isolation and harsh conditions in the heavily forested areas where much back-breaking labour was required to clear the way for both habitation and travel — one of the reasons why travel by water was generally preferred to that over land. – NAC: C-11811

1.6 Chatham, Upper Canada, 1838: Another view by Bainbrigge also portrays conditions which could have been seen in many a pioneer settlement throughout the Niagara Peninsula: the tree stumps still along the road, timber houses, and probably a shop at the left, a substantial stone dwelling, split rail fences and a simple cart. – NAC: C-11878

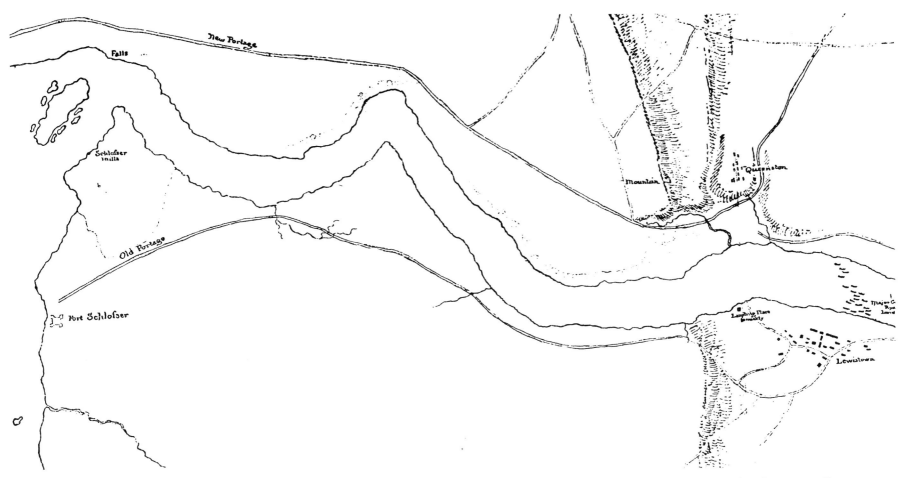

1.7 The Portage Route Between Queenston and Chippawa, 1814: While the spectacular Niagara Falls have been a much-visited and romantically described attraction for centuries, they were a serious impediment to pioneer travel, necessitating an awkward and costly portage. The French and Indians had long had a portage route along the east side of the river, but following the American Revolution the British felt the need of a "safe" route on their side of the river, hence the "New Portage" as opposed to the "Old," as shown on this military map. – NAC: NMC-18466

1.8 Chippawa (or Fort William), ca. 1805: George Heriot (1759-1839) was another British artist who portrayed early Canadian life. Heriot served in the colonies as a civil servant (head of the postal service, 1799-1816). He travelled widely, and in 1807 published *Travels Through the Canadas*, in which this view of the wooden bridge carrying the road to Fort Erie appeared. Along this road supplies were conveyed by land over the portage road to and from Queenston. At Chippawa the supplies were loaded on ships for transport to Fort Erie. Heriot describes the village thus: "There are . . . some mercantile store-houses, and two or three taverns." – McCord Museum of Canadian History, Montreal

The mouth of this creek forms a very fine scene; a very bold spur of the high land appears beautiful in the distance. It is about three miles off. Some cottages are pretty placed [sic] on the banks of the river, and a sawmill affords a quantity of boards, which, piled up in a wood, makes a varied foreground.
– [Simcoe *Diary*, p. 222, 10 May 1794]

1.9 Mouth of the Forty Mile Creek, 1790s (above): Another of Mrs. Simcoe's delightful sketches provides a glimpse of a tiny community on the south shore of Lake Ontario (40 miles from the Niagara River). Such settlements were located on the creeks — the Twelve, the Twenty and the Forty — either near the mouth or further upstream, where a fall of water could provide power for a mill. (She also shows the travel accommodation the governor's party could command.)
– AO: Simcoe 83E

1.11 "View of Fort Erie with Migration of Wild Pigeons," 1804: Lt. Surgeon Edward Walsh (1756-1832) produced this watercolour about the time George Heriot described this small link in the British defence system as follows: "The old fort on the west side of the entrance into the lake, consists of no more than a few houses, a block-house of logs, with some habitations for commercial people, and one or two store-houses. A new stone fort, in the form of a quadrangle, is now constructing on rising ground behind the block-house. A company of soldiers is usually stationed here, and the men are chiefly employed in assisting to conduct the transport of stores. Two vessels in the service of the British government are used in navigating the lake [*Travels Through the Canadas . . .*, p. 174]." – ROM: 952.218

1.10 "The Joys of River Travel, North America," ca. 1806 (opposite): Such might have been the title given to this watercolour by an American artist. Experiences like this were common on Canadian streams. On 28 July 1794 Mrs. Simcoe recorded: "We were no sooner in the boat, expecting a rapid passage up the Twenty-Mile Creek, when the wind veered and came right ahead, so that it was ten o'clock before we arrived at the inlet. It was quite dark, and we were another hour getting the boat over the sand, and rowing to the house [*Diary*, p. 234]." – Mansell Collection

1.12 Launching of the NEWASH and the TECUMSEH, 1815: This sketch by an anonymous artist is not necessarily accurate, but records a dual launching of ships at Chippawa Creek. No precise location is specified, but possibly it was at the village of Chippawa, where we know a shipyard existed by 1832 [see Fig. 1.13]. – ROM: 967.106.2

1.14 Balls Falls, Twenty Mile Creek, ca. 1815 (opposite): The Niagara Escarpment impeded travel along the streams which flowed over it, but these same streams powered the water wheels which ran the grist and saw mills vital to pioneer communities. George Ball's grist mill, with a 9.8 m (32-ft) water wheel, began operation in 1810. Unfortunately the water flow in these streams was subject to seasonal variations: too much too suddenly during spring freshets, too little in the dry summer months. This is why a group of mill-owners in the St. Catharines area supported William Hamilton Merritt when he proposed to build a canal between the Chippawa Creek (Welland River) and Lake Ontario to ensure a steady water supply and to transport boats between lakes Ontario and Erie via the Welland and Niagara rivers. – ROM: 951.41.12

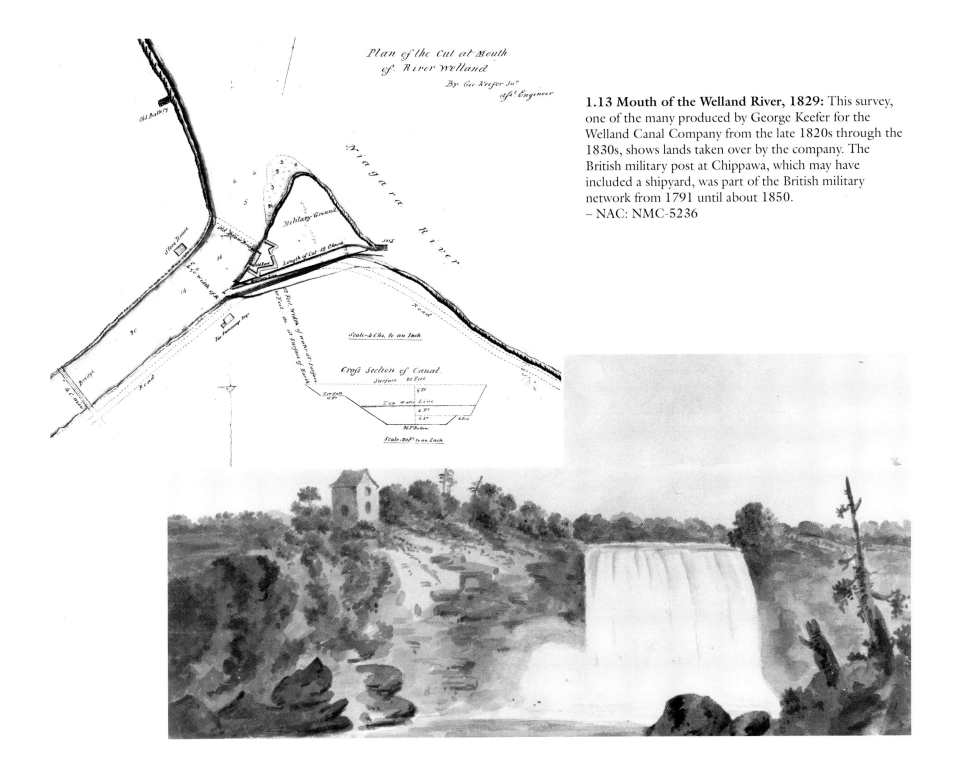

Plan of the Cut at Mouth
of River Welland

By Geo Keefer Jnr
Asst Engineer

Niagara River

1.13 Mouth of the Welland River, 1829: This survey, one of the many produced by George Keefer for the Welland Canal Company from the late 1820s through the 1830s, shows lands taken over by the company. The British military post at Chippawa, which may have included a shipyard, was part of the British military network from 1791 until about 1850.
– NAC: NMC-5236

1.15 The Road Along the Mohawk River, N.Y., 1828: Many of the conditions which prevailed in Upper Canada were equally prevalent in New York State. While the road shown here looks far better than some of the early roads in the Niagara Peninsula, the need for a reliable water route, which would provide steady water supplies for mills and avoid the series of rapids along the river's course, was strong and led to the beginning of construction of the Erie Canal in 1817. – Library of Congress

1.16 The Harbour at Buffalo, N.Y., 1827 (opposite): This sketch by Capt. Basil Hall indicates that the Erie Canal (opened 1825), in its first decade of operation, was bringing prosperity to American ports. The men who proposed building a canal to link lakes Ontario and Erie in British territory were well aware of the danger to the economy of the Canadas that the American canal posed. It threatened to take all trade from the Great Lakes area through American territory to the Atlantic seaboard at New York and thence abroad. – SCHM: N-4605

In the light of conditions such as those described above it is not surprising that by the 1820s support for a canal in British territory came from many sources, including both local farmers and men of influence in Lower Canada:

Hear is a canal begun in this part of the countrey caled the Welland Canal to join the two lakes together. It is intended to cut of the land cariage over the Falls of Neiagara, which, if compleeted, would be of great servis to this countrey. It is not sertainly know the rout that it is to go from the Welland to the Grand River. It is expected to go near this place; perhaps it may go through our farme. I feel willing that it might. It is in agetation to have it for a sloop navegation. I could wish to see the same compleeted, then if aney of you sho7uld have a minde to come you might come all the way by water conveyance. [Thomas Priestman (who had emigrated to America about 1806-7, and settled in Wainfleet in 1811) writing to his brother John, still in England, 27 March 1825. Quoted in Angus J.L. Winchester, "Scratching Along Among the Stumps," *Ontario History*, Vol. LXXXI, No. 1, March 1989, p. 51.]

3. Resolved. — As the opinion of this meeting that the completion of the said Canal is an object of the first importance and would be productive of the greatest benefit and advantage to the commerce of the Country not less to this Province than to that of Upper Canada.

4. Resolved. — As the opinion of the meeting that the said undertaking merits every encouragement, and the aid of every inhabitant of this Province who feels an interest in the agricultural and commercial prosperity of the Canadas. [Meeting of Merchants and other inhabitants of Quebec City, 11 March 1824, quoted in E.A. Cruikshank, "The Inception of the Welland Canal," *Ontario Historical Society Papers and Records*, Vol. XXII (1925), p. 77.]

Chapter 2: Merritt's Crazy Crotchet

2.1 Italian Canal Scene, 17th Century: When William Hamilton Merritt and some of his associates in the Niagara Peninsula in the 1820s envisioned a canal to connect lakes Ontario and Erie, they knew of precedents in both North America and Europe. In the 17th and 18th centuries, Europeans gradually developed two devices which permitted the construction of man-made waterways to overcome natural obstacles: the lock (illustrated here), to raise or lower boats from one level to another, and the aqueduct, to carry the channel across a valley. Note the tiled roof over the lower lock gates (open to allow the boats to exit) and the circular shape. Otherwise, this Italian drawing shows all the elements of modern locks, albeit in most elementary form. – New York Public Library

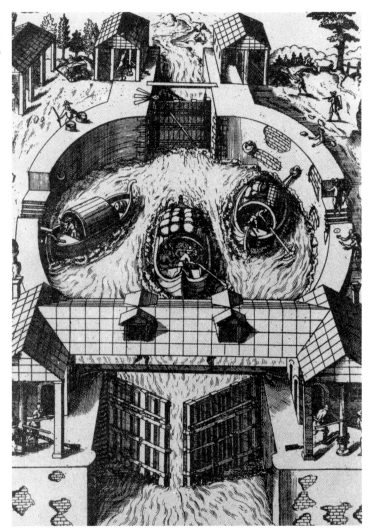

2.2 French Canal Scene, 18th Century: The French encyclopedist, Denis Diderot (1713-1784), included this drawing in *L'Enyclopaedie ou Dictionnaire Raisonné des Sciences, des Arts et des Metiers*, published from 1751 to 1772. Such drawings, as well as those in a section on canals in the American *Cyclopedia* published by Rees from 1810 to 1824, were referred to by early canal builders in North America. – Brock University Library

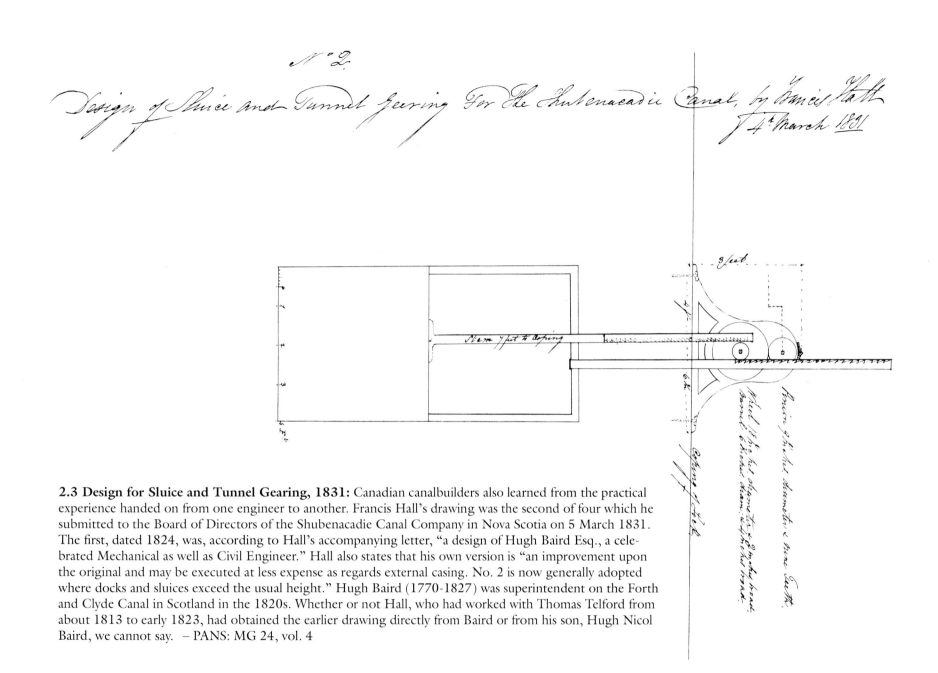

2.3 Design for Sluice and Tunnel Gearing, 1831: Canadian canalbuilders also learned from the practical experience handed on from one engineer to another. Francis Hall's drawing was the second of four which he submitted to the Board of Directors of the Shubenacadie Canal Company in Nova Scotia on 5 March 1831. The first, dated 1824, was, according to Hall's accompanying letter, "a design of Hugh Baird Esq., a celebrated Mechanical as well as Civil Engineer." Hall also states that his own version is "an improvement upon the original and may be executed at less expense as regards external casing. No. 2 is now generally adopted where docks and sluices exceed the usual height." Hugh Baird (1770-1827) was superintendent on the Forth and Clyde Canal in Scotland in the 1820s. Whether or not Hall, who had worked with Thomas Telford from about 1813 to early 1823, had obtained the earlier drawing directly from Baird or from his son, Hugh Nicol Baird, we cannot say. – PANS: MG 24, vol. 4

2.4 Stump-Pulling Machine, Erie Canal, 1817-25: Many of the American engineers and contractors, who had taught themselves the skills needed to construct the Erie, and who had also designed and built equipment necessary for specific tasks (as with this gigantic device), brought their expertise to help build the first Welland Canal. Without their knowledge, and that of Francis Hall, the task of transforming the dream of a canal in the Niagara Peninsula into a reality would have been far more difficult, if not impossible. – R.M. Styran

2.5 Site of the Merritt Mill on Twelve Mile Creek, 1827: Sheriff Thomas Merritt constructed a saw-mill in 1816, in what is now known as the Welland Vale area of St. Catharines. Since the water supply was never satisfactory, Thomas's son, William Hamilton Merritt, seized upon the idea of a waterway connecting the Welland River to the Twelve Mile Creek, which would assure a more reliable flow, permitting more efficient use of his mill. Other mill-owners in the area heartily endorsed the idea, hoping for similar benefits. – Board of Public Works, *Register B, 1841-1880*

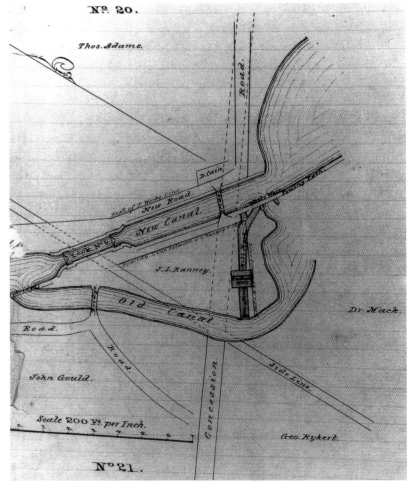

2.6 Proposed Routes of the Welland Canal:
From 1818 until 1824, when construction began, and indeed until the Feeder was built in 1829, the proposed route for Merritt's canal generated considerable, and at times heated, debate. In 1823-24 two main possibilities were surveyed, resulting in the farmers and businessmen in the different settlements arguing for, or against, the route that seemed most beneficial, or harmful, to their interests. – J.N. Jackson, *St. Catharines, Ontario: Its Early Years.* Mika, 1976, p. 189

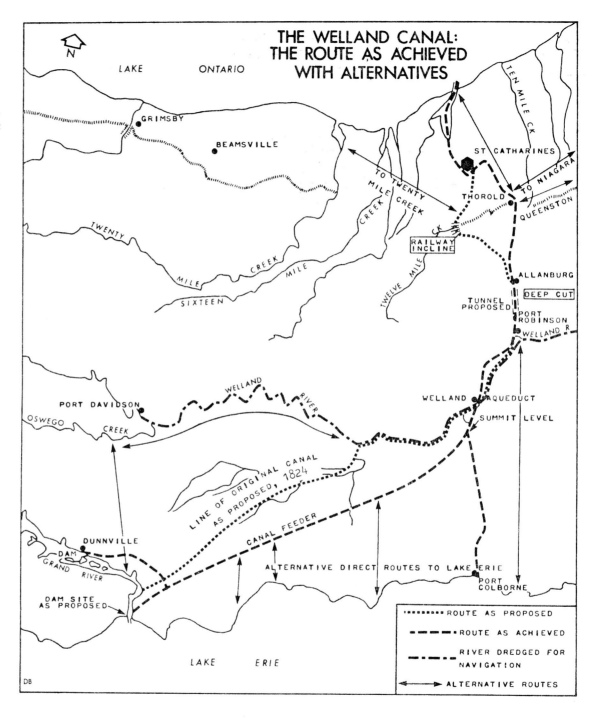

From a Drawing by W.R. Callington Engineer, Boston, from an Actual Survey made in 1837.

A Birds eye View of the river Niagara from Lake Erie to Lake Ontario shewing the situation and extent of NAVY ISLAND, and the Towns and Villages on the banks of the river in Canada and the United States._ with the situation of the Caroline Steam Boat off Schlosser.

UNITED STATES.		CANADA.	
1 Town of Buffalo	6 Hotel at the falls	A Lake Erie	F Rapids
2 Black Rock	7 London	B Fort Erie	G Cataracts of Niagara
3 Grand Island	8 Fort Niagara	C Waterloo	H Queens Town
4 Tonewanta Creek	9 Lake Ontario	D Navy Island	I Fort George
5 Grand Canal	10 Goat Island	E Chippewa	J Welland Canal
	11 Schlosser, the place where the Caroline was burnt.		

2.7 Niagara River, 1837: This bird's-eye broadside, lithographed and published in London, England, was taken from a survey made by W.R. Callington, an engineer from Boston. The artist's chief interest was not commerce, but the military and naval positions on both sides of the river at the time of the rebellion by William Lyon Mackenzie. It is obviously based on an earlier map, since the extension of the Canal to Lake Erie, completed in 1833, is not shown. However, the drawing does show quite clearly that the Welland Canal was a response to both the problem of the Escarpment and the possible attack from the United States, providing a means of transport between the lakes and well away from the frontier.
– Metropolitan Toronto Reference Library: T-14981

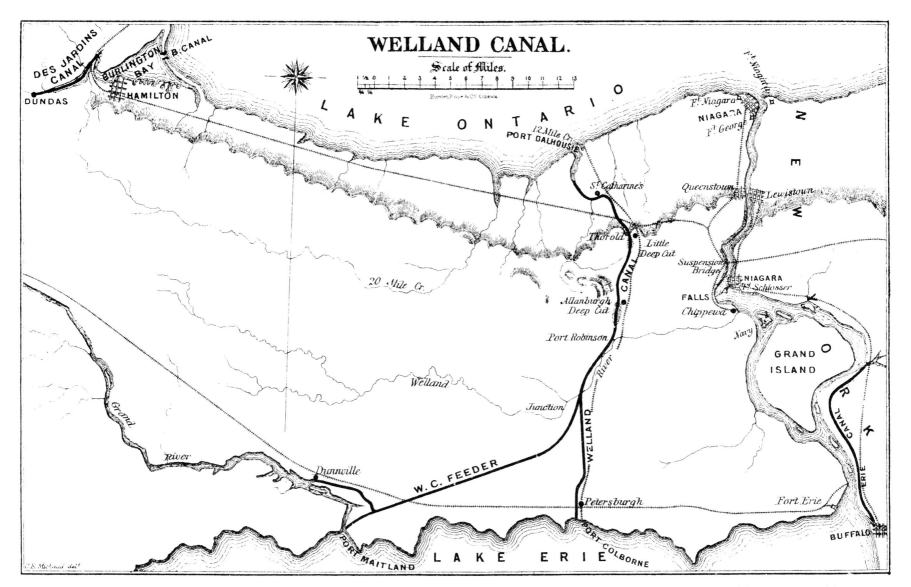

WELLAND CANAL.

Scale of Miles.

Hunter, Rose & Co. Ottawa.

2.8 Niagara Peninsula, ca. 1860: By this time military danger from the United States was negligible, and the economic threat had been countered by construction of the Welland Canal and the Welland Railway (opened 1859). Both the Erie and Welland canals had been enlarged and improved, and the thriving communities spawned by them on both sides of the border more than justified the vision and faith of their proponents, New York State Governor DeWitt Clinton and St. Catharines businessman and entrepreneur William Hamilton Merritt [see Fig. 2.9, The Portrait Gallery]. – NAC: NMC-117641

DeWitt Clinton
Governor of New York State
– New-York Historical Society
New York City

J. Barentse Yates
New York Financier
– Village Hall, Chittenango, NY

William Lyon Mackenzie
Newspaper Publisher
Toronto and Niagara
– Metropolitan Toronto Library:
JRR-363

Hamilton Killaly
Irish Civil Engineer
– Metropolitan Toronto Library:
JRR-788

George Ramsay
9th Earl of Dalhousie
Governor-in-Chief of the Canadas,
1819 to 1828
– Metropolitan Toronto Library:
T-31639

Thomas Coltrin Keefer
Canadian Civil Engineer
– NAC: PA-011852

James Geddes
American Civil Engineer
– New York Society Library

Samuel Keefer
Canadian Civil Engineer
– NAC: C-21683

William Hamilton Merritt
St. Catharines Entrepreneur
– Page's *Historical Atlas for the
Counties of Lincoln and Welland*

Paul Shipman
United Empire Loyalist
St. Catharines' Tavern Keeper
– Metropolitan Toronto Library:
T-30363

Hugh Nicol Baird
Scottish Civil Engineer
– James T. Angus, A
Respectable Ditch. McGill-
Queens Press, 1988

2.9 THE PORTRAIT GALLERY

While it took men with practical knowledge of construction methods to build the canal, without men of vision (and political and economic influence) the project might have been stillborn. **William Hamilton Merritt** (1793-1862) was the instigator of the first Welland Canal, but he had support at both the local level (**Paul Shipman** (1756-1825, for example), and at the apex of the colonial political establishment, **Lord Dalhousie** (1770-1838). Merritt also had the benefit of an example, the Erie Canal, which he visited on many occasions. He consulted with the driving force behind the Erie, New York State Governor **DeWitt Clinton** (1769-1828) and he obtained American financial backing with the assistance of **J. Barentse Yates** (1784-1836), a prominent New York financier. In fact it was at the urging of Yates and other New York businessmen interested in buying stock in the Welland Canal Company that the original plan for a barge canal (similar to the Erie) was changed so that the Welland could accommodate larger vessels, even steamships, since these men had the vision to see that steam would prevail.

Newspaper publisher and publicist (later political agitator and rebel) **William Lyon Mackenzie** (1795-1861) helped promote the venture with fulsome praise in the early stages, although by 1836 he was harshly critical of Merritt's involvement in the Welland Canal Company and its affairs. A lengthy investigation by a committee of the House of Assembly of Upper Canada eventually concluded that while "the books have been kept in a very careless, irregular and improper manner, highly discreditable to a public body . . . your committee cannot say that any intentional fraud against the public or canal proprietors, has been brought home to any individual officer of the Company . . . " [*Report*, p. 4]. The following year, a frustrated Mackenzie led a short-lived rebellion against the government.

Among the practical men hired to produce surveys in the early stages of the project were James and Samuel Clowes (Englishmen; Samuel did a survey for the Rideau in 1824), and Hiram Tibbets, an American. Other Erie veterans included Benjamin Wright (1770-1842), who gave a hearty recommendation to his assistant, Nathan S. Roberts [see Introduction, Fig. Int. 12]; Alfred Barrett, resident engineer from 1826 until 1828, again briefly in the early 1830s, and again from 1848 until his death the following year; and **James Geddes** (1763-1838), who worked with Wright on the Erie and who was brought in to help Barrett re-survey the line following the collapse of the Deep Cut banks in 1828.

Francis Hall [see Figs. 2.3 and Introduction, Fig. Int. 11] was hired in the early stages to assess surveys and estimates of James and Samuel Clowes, and again in 1836, when the canal had fallen into grave disrepair. At that time he supervised a major rebuilding. In May 1837 Scottish-born **Hugh Nicol Baird** (1796-1849) and Irish-born **Hamilton Hartley Killaly** (1800-74) were employed to survey both the current state of the canal and a possible route for a proposed enlargement, which became the Second Canal. Baird turned down an appointment as permanent engineer in charge of the Welland, but Killaly accepted the post, remaining as engineer from 1838 until he was appointed chairman of the newly formed Board of Works of Upper Canada early in 1840. By that time, **Thomas Coltrin Keefer** (1821-1915) and **Samuel Keefer** (1811-90), two sons of George (first President of the Welland Canal Company), had become the first generation of Canadian-born engineers, and both served on the Welland (Samuel in the 1830s and again in the 1840s, and Thomas in the early 1840s). Both went on to other civil engineering projects. A fourth-generation engineer, another Thomas Coltrin Keefer, was active in the middle years of the 20th century.

All of these men, in their varying capacities, helped William Hamilton Merritt to realize his "crazy crochet" — the first Welland Canal.

Chapter 3: Milling and Manufacturing

3.1 St. Catharines, ca. 1920: From its inception the Welland Canal has attracted industry. Lock 4 of the Second Canal and the three levels of the hydraulic raceway are all clearly visible. A number of the industries originally drawn to the area by both the waterpower available from the raceways and the ease of access to transportation remained even after the routes of the Third and Fourth canals by-passed the city. The four-storey structure (centre right) was originally a flour mill associated with William Hamilton Merritt and the Phelps family [see also Chap. 4, Fig. 4.4 and Chap. 12, Fig. 12.7]. – SCHM: N-8366

3.2 Near Port Colborne, ca. 1880: One of the earliest grist mills in the Niagara Peninsula was built near Sugar Loaf Hill in 1788 by Christian Zavitz. It is typical of the pioneer era: individually owned and operated as a "custom" mill (grinding flour to order for neighbouring farmers) as opposed to a "merchant" mill producing large quantities of flour for export. Built of stone as far as the top of the water-wheel, the upper storeys are of wood. We have no photographs of similar mills along the course of the Canal, since most were re-built and enlarged several times.
– PCHMM: P-979.952

3.3 Regulations for Granting of Mill Sites, 1852:

In the granting of "privileges" [i.e., for mill sites], preference will be given in the following order:

1st. To Manufacturories requiring the aid of expensive machinery, and the employment of considerable labour.
2nd. To Grist Mills.
3rd. To Carding and Fulling Mills, &c.
4th. For Mechanical purposes, such as planing, turning; pail, last, wainscot, and sash-making, &c.
5th. Saw Mills.

The water in all cases to be applied through the medium of driving-wheels of the most approved modern principle, as to small consumption of water. The regulating weir and gates, for the discharge of the water through the canal bank, to be constructed by the Department of Public Works — six per cent. on the cost of which is to be paid by the tenant, in addition to his rent. – Wm. H. Smith, *Smith's Canadian Gazetteer.* Toronto, 1849

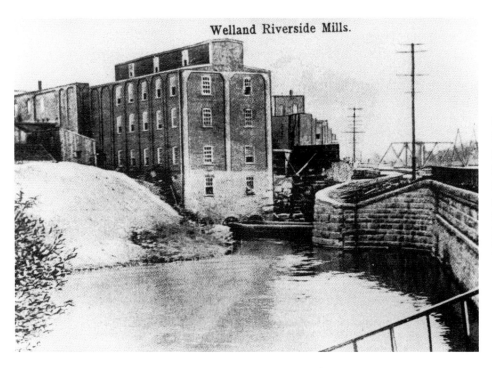

Welland Riverside Mills.

3.4 Riverside Mills, Welland, 1907: Established by Robert Cooper in 1892, this handsome brick structure was much larger than earlier mills, but was still water-powered. It was the last large flour mill built along the Welland Canal before Robin Hood Mills (1940) at Port Colborne. Flour milling had become concen trated at the southern terminus of the canal earlier in the 20th century, following construction of the government (1907) and Maple Leaf Mills (1908) elevators [see Chap. 9, Fig. 9.12]. – Welland Public Library

3.5 McCleary & McLean, Merritton, 1879: In addition to flour, pioneer society required another basic product, lumber for construction. Although the first sawmills were erected on streams, millers were soon attracted to the banks of the canal, with its more reliable water supply. In the 1840s McCleary and McLean set up their mill on the Second Canal. They supplied not only building materials for private and commercial use, but also nautical timber for the flourishing shipyards along the canal [see Chap. 5, Fig. 5.3a]. – Evans' *Gazetteer and Business Directory . . . , 1879*

3.6 S. Reichmann and Son's Planing Mill, Humberstone, ca. 1890: This was another of the many mills which supplied construction materials both for the growing canal-side communities and for export during the second half of the 19th century. All the staff (and presumably Mr. Reichmann's family) have been lined up for the photographer, but who is on the roof, and why? – PCHMM: P-980.8

3.7 Shickluna's Shipyard, St. Catharines, 1894: By this time Shickluna's yard, one of the earliest shipbuilding establishments on the Dick's Creek section of the canal, was in the hands of Frank Dixon and son. The tugs CHARLES E. ARMSTRONG and JESSIE HUME are under construction here. The first shipbuilder in the St. Catharines area was Russell Armington, who had sold his shipyard outside Troy, N.Y., about 1825 and set up business on the 12 Mile Creek in 1827. Following his death in 1837 the yard was taken over by the Maltese shipbuilder Louis Shickluna, who operated both it and the dry dock on Dick's Creek until his death in 1880. From 1838 to 1880 Shickluna built over 100 vessels of various kinds. The Shickluna yard was only one — if perhaps the best known — of its kind along the Welland Canal, and R. Soper was but one of many men engaged in providing sails and other equipment for ships which plied the waterway. As with other industries [see Fig. 5.5, for example, Soper also served the communities along the canal. – SCHM: N-2794; inset *St. Catharines Directory . . . for 1881-82*

3.8 Port Colborne, ca. 1873: For most of the 19th century and into the early years of the 20th century, wooden sailing ships and steamers were constructed in most of the canal communities. While Port Colborne was never an important shipbuilding centre, a number of tugs originated there. The three gentlemen staring at the camera are standing on a tug under construction, possibly the HECTOR, built in 1873. Prominent in the background is the Grand Trunk Railway grain elevator, built in 1860. – PCHMM: P-987.2

3.9 Port Weller, 1989: Today the only shipyard on the Welland Canal is the Port Weller Dry Docks, sole representative of a long and rich heritage. Recently the HMCS NIPIGON underwent a complete overhaul and refitting, including accommodation for 47 women, thus becoming the first "co-ed" fighting ship in the Canadian navy. Originally it was intended that Port Weller, as the Lake Ontario terminus of the Fourth Canal, should become a thriving community. The Depression of the 1930s, combined with the fact that it was not on major road or rail networks, frustrated that hope, and today the dry docks are the only major employer. – R.M. Styran

3.10 Port Maitland, 1989: Shipyards were also located in some of the Feeder Canal communities, such as Dunnville, Port Maitland and Stromness. Powell's Shipyard is the now the only shipbuilder in the area. Founded in 1946 by S.G. Powell in Scottsville, later located at a site near the Dunnville dam, the yard was moved to its present site by the founder's son in 1986. The building and repair of commercial and pleasure boats has been facilitated by a new floating dry dock, recycled from the gate lifter used in construction of the Fourth Canal. – R.M. Styran

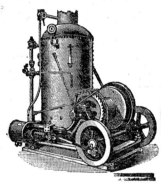

3.11 Dobbie & Stuart, Thorold, 1892: A relative newcomer to the iron foundry scene along the canal, this firm, founded in 1865 as Thorold's first foundry, survived on the same premises into the recent past [see Chap. 11, Figs. 11.16-18]. Foundries were another necessity of pioneer life, since iron was required for agricultural implements, and pumps and engines of all kinds. Some of the foundries added machine shops specializing in a wide variety of products. – *Ontario Gazetteer and Directory . . . 1892-93*. Toronto, 1892

3.12 Oill's Foundry, St. Catharines, 1863: The machine shop and foundry of George N. Oill, on St. Paul Street near James, was established in 1850. It became prominent in the area, manufacturing mowing machines, steam engines and boilers, various agricultural implements, and marine engines. The Oill(e) family was another of the second generation of entrepreneurs who contributed to the development of St. Catharines as an important regional service centre. George N. Oill and his brother Ezra built the first threshing machine in the district, and another brother, Jerome, was also involved in the foundry business. Jerome was married to a granddaughter of Oliver Phelps, contractor on the first Welland Canal. The youngest brother, Lucius Sterne Oille, became a well-known and loved physician, and served as mayor. He was involved in setting up the St. Catharines Street Railway and the town's waterworks. – Fuller's *St. Catharines Directory for 1863-64*. St. Catharines, 1863

WELLAND
STEAM BRICK AND TILE WORKS
HOOKER & SONS, Proprietors.
Manufacturers of the best description of
BRICK AND TILE
At the Lowest Market Rates,

WELLAND. - - ONTARIO.

☞ All Orders Promptly Attended to.

3.13 Hooker Brick Yards, Welland, 1907: Founded in 1855, Hooker's provided bricks for many buildings still standing in Welland, such as that occupied by Bogner's Photography. The company benefitted from excavations for the Third Canal, which provided an excellent clay, giving a distinctive bright-orange brick. – Wm. W. Evans, *Gazetteer and Business Directory for Lincoln and Welland Counties for 1879*. Brantford, 1879

3.14 Union Carbide, Welland, 1990: Typical of a newer generation of chemical and metal industries established in Welland in the early years of the 20th century, Union Carbide (1907) is located on Canal Bank Road. Such industries require large amounts of water for cooling and washing purposes, hence still make use of canal water, although they are not necessarily dependent on the canal for transportation. – R.R. Taylor

3.15 Lybster Cotton Mill, Merritton, ca. 1900: While pioneer households produced wool and linen for family clothing, textile mills were needed for carding and spinning of wool, and for the production of blankets and woollen cloth. The first woollen mill in Upper Canada was established in St. Johns [see Chap. 1, Fig. 1.1] in 1813. Lybster's, founded in 1862, on the Second Canal, employed 200 hands in 1869, working on 261 looms and 11,500 spindles. The premises became part of the Lincoln, then Alliance, later Domtar, paper mill. – SCHM: N-8502

3.16 Canada Hair Cloth, St. Catharines, 1990: Canada Hair Cloth was not established until 1884, moving to its present site in 1888. The move to the Second Canal site, originally occupied by the Dolphin Paint Company, was made in order to utilize the raceway's water power, even though by then the Second Canal carried less trade than the new Third Canal, which was already in operation. Unlike many canal side industries, the company is still in production on the same site, having undergone expansion in 1911 and 1915. – R.R.Taylor

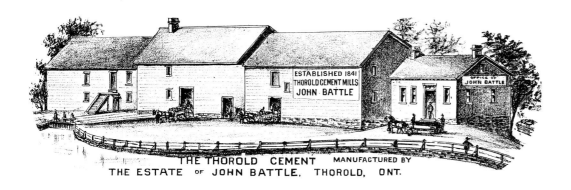

3.17 Battle's Cement Works, Thorold, ca. 1890: This establishment dated back to the last days of the First Canal, and produced mortar and cement foundations, floors, etc., for many of the brick and stone houses, factories and commercial premises which benefitted from construction of the Second Canal from 1842 to 1845. Their hydraulic cement was used in many of the major construction projects of the late 19th century: the Aberdeen Bridge and the Grand Trunk Railway Bridge at Oakville, the Victoria Tubular Bridge in Montreal, the tunnel under the St. Clair River, and both the Second and Third Canals. – St. Catharines Centennial Library, Photo Album 13, no. 25

3.18 Woodburn "Sarven" Wheel, St. Catharines, ca. 1890: Another of the local industries featured on a promotional bird's-eye map of about 1890 [see also Chap. 5, Fig. 5.3], this company was established in 1862 on the Second Canal. It used both the canal's water power and its access to transportation – raw materials brought in and finished products shipped out. This was just one of the companies specializing in wood fittings for carriages and in wheels of all types. Some of these companies went on to manufacture steel wheels and helped lead St. Catharines into the manufacturing of automobile parts. – St. Catharines Centennial Library, Photo Album 13, no. 21

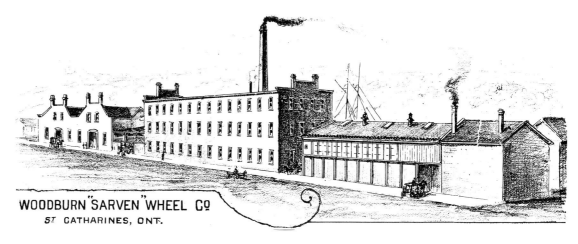

3.19 Lincoln Paper Mills, Merritton, ca. 1890: Built in 1877, the plant produced manila paper, newsprint, grocery bags, and flour sacks. Mill 'A' was on the raceway; Mill 'B' on the Second Canal itself. Mill 'A' has been enlarged and has changed hands several times. Part of it is now used as a power plant for Domtar Fine Papers. Following developments in the 1860s, including the use of ground wood instead of rags to produce paper, other paper mills were established in the Merritton-Thorold area. Some were on the site of, or near to, lumber mills, where a plentiful supply of ground wood was available. There are now four major paper companies in this area: Kimberly-Clark, Domtar, Quebec and Ontario Paper (formerly Ontario Paper), and Fraser (formerly Provincial). – St. Catharines Centennial Library, Photo Album 13, no. 24

3.20 Noah A. Phelps: A great-nephew of the First Welland Canal contractor, Oliver Phelps, Noah (ca. 1825-1900) was part of a dynasty of entrepreneurs in the St. Catharines-Thorold area. They ran a flour mill and sawmill, a spoke factory and a hammer works, and established comfortable homes, some of which can still be seen along the route of the Second Canal. Noah was also involved with the Lincoln Paper Mills [see Fig. 3.19]. – SCHM: N-4063

3.21 Cronmiller & White's Brewery, Port Colborne, 1892 (opposite): Henry Cronmiller, of Alsatian descent, began his career as a farmer, then in about 1860 added a store and hotel in Humberstone to his interests. Two years later he sold all of these, moved to Port Colborne, and began buying grain. In 1875 he established his brewery, and this handsome building became a visual asset to the community, while providing good brews for both local quaffing and distribution further afield. Cronmiller is typical of many of the entrepreneurs who established themselves in a small way in one of the canal-side communities, then became involved in regional and national trade. – PCHMM: P-979.547; inset PCHMM: P-979.550

Henry Cronmiller

1892

Chapter 4: Trade and Commerce

4.1 St. Catharines, ca. 1890 (opposite): The political and financial support given by merchants in Lower Canada to Merritt's project was based on their expectations of increased trade with the interior. At St. Catharines, for example, ships could turn in at the rear of Sylvester Neelon's mill at Lock 4 to take on flour for Montreal. Barrels were made in a cooperage owned by the mill (at right), and the large building between the masts was where the empty barrels were stored until needed [see Chap. 3, Fig. 3.1]. The steamer OCEAN (centre), owned by Neelon, took a load of flour to Montreal once a week throughout the shipping season. Both the OCEAN and the propeller EUROPE (left) were typical of ships plying the Canal in the late 1800s. In 1895 Packard Electric took over the stone mill [see also Chap. 12, Fig. 12.7] – SCHM: 982.207.4

4.2 and 3 Trade designs, 1850s: The symbolic and romantic figures of the sailor, with barrel and bale to indicate produce, and Prosperity, with her flowing cornucopia, appears on a promissory note of 17 April 1857. A more realistic representation of an important aspect of 19th century trade is the drawing of the weigh scales, which also shows how the pervasive barrels were identified. The gentleman may be a shipper or an official responsible for the verifying of weights. – AO: Norris and Neelon Papers, MS 490.1-2-1; Lovell's *The Canada Directory . . . 1851*

4.4 St. Catharines, 1858: This shipping order indicates that the trade with Montreal was not entirely free from difficulties. The note at the bottom reads: "If the St. Lawrence Canals are impassible the above flour is to be delivered into the "Grand Trunk R. Road" at Prescott they paying Lake Freight at the rate of 15¢ fifteen cents per Barrell and to be forwarded by them unto Mssrs. Macdougal & Budden Montreal." – SCHM: N-4594

4.5 Ship's Manifest, Montreal, 1857 (opposite): On the return trip from Montreal, the Norris and Neelon schooner PETREL would bring back goods perhaps not available in the Niagara Peninsula, such as the iron, steel, chain, nails and soda ash listed here. Newspapers and directories of the period carried numerous advertisements inserted by local merchants and proclaiming the advantages of the various imported goods which they had available: groceries, wines and liquors, clothing, furniture and furnishings of all kinds the list is endless [see also Chap. 5]. – SCHM: Norris Papers

No. 12   Montreal, 29ᵗʰ October 1857

Shipped, in good order and well conditioned, by **JOHN MACPHERSON & Co.**, as Agents and Forwarders, for account and risk of whom it may concern, on board the Schr. Peter D. Graham Master, the following articles, marked and numbered, as in the margin, and to be delivered in like good order and condition (the leakage of Oil, Molasses, Liquors, and other Liquids, the dangers of Navigation and Fire excepted,) unto the Consignee named in the margin, or to _____ Assigns; freight and charges to be paid as noted below.

In Testimony Whereof, the Master of said vessel hath signed two Bills of Lading of this tenor and date.

	MARKS.	No. PC'S.	DESCRIPTIONS.	WEIGHT.	RATE.	AMOUNT.	TOTAL.
Thomas McGevins Thorold		223	Bars Iron				
			4 Bdle do				
			2 Bdle Steel				
			2 Keg Chains				4 5 9
Henderhot & Schwaller Thorold	H·S	14	Kegs Nails				2 3
Wm B. Henderhot Thorold		5	Casks Soda ash	65			2 6 11
						Codbt £	7 14 11
						7 14 11	
						4 14 8	

55

4.6 and 7 Provisions Trade, 1870s and 1880s: These firms, like many others in the canal communities, owed their prosperity (if not their very existence) to the Welland Canal. J.T. McComb of Merritton was typical in that he not only imported goods from Montreal, but also supplied the ships as they passed through the canal; and one may assume that some of the goods indicated on Hendershot's bill had also come in by ship. – E.H. Williams, *Directory of St. Catharines, Thorold, Merritton and Port Dalhousie, for 1877-78*, St. Catharines, 1877; SCHM: Norris Papers

J. T. McCOMB,

GROCER & PROVISION DEALER.

(Fancy and Small Wares.)

—

Special attention to SHIP SUPPLIES.

—

Cor. Thorold and Concession Roads,

OPPOSITE LOCK 7.

Thorold, *August 25* 1887.

A. Schooner Louisa)

Bo't of **Hendershot & Moore,**

DEALERS IN DRY GOODS, GROCERIES, AND PROVISIONS,

CORNER ALBERT & CANAL ST., (Opposite Lock No. 24.)

50 lb	Ham	16	$8.00
100 "	Flour	4½	4.50
1½ Bush	potatoes	7/	1.31
5 "	Butter	1/8	1.04
2 Doz	Eggs	20	0.40
			$15.25

Rec'd payment

Hendershot & Moore

4.8 and 9 The Timber Trade, 1874 and 1953: (opposite) Irish-born William Beatty, like many other merchants, carried on a profitable timber and lumber business. As with the provisions merchants, his trade was directed both at the private and commercial firms in the communities, and at the canal traffic. The YANKCANUCK (1) represents a long line of ships carrying timber and lumber through the Welland Canal. Built in Detroit in 1889 as the bulk freighter MANCHESTER, it underwent two rebuildings in the 1920s and served various owners until 1959 — a very long career indeed. It was the last of the "composite" ships (built of oak timber and charcoal iron) to work on the Great Lakes. – J. Horwitz, *St. Catharines General and Business Directory for 1874*, St. Catharines, 1874; *Inland Seas*, Vol. 15, 1959, 9. 126

4.10 Ontario Paper Mill, Thorold, 1931: The *Chicago Tribune* is the parent of Ontario Paper, now Quebec and Ontario Paper (established 1912). This plant, located on the Fourth Canal and the Canadian National Railway, was capable of an output of 400 tons per day in 1931, all of which was destined to pass through the canal on its way to Chicago. The THOROLD, built in England in 1930, had a raised trunk deck to allow maximum loading of newsprint (68,562 tons). She was lengthened at Port Weller in 1962, continued to serve on the Great Lakes until 1986 (carrying grain and other cargo after paper was transferred to the railroads in the 1970s), and was scrapped early in 1989. – St. Lawrence Seaway Authority: Western Region

ONTARIO PAPER COMPANY LIMITED

3000 TONS OF PAPER *for* THE CHICAGO TRIBUNE

EXPORTS FROM THE PORT OF DUNNVILLE, FOR THE YEAR 1850

SHIPPED TO FOREIGN PORTS

Flour	12,910 barrels
Wheat	51,948 bushels
Oats, Barley, etc.	10,960 bushels
Oatmeal	200 barrels
Gypsum, unground	1,915 tons
Potash	39 tons
Limestone	70 tons
Square timber, in rafts	88,000 cubic feet
Flatted and rounded timber	31,000 feet
Pine lumber	13,555,000 feet
Barrel Staves	10,000 No.
West India Staves	18,000 No.
Shingles	1,340 M.
Saw Logs	1,340 No.
Lath, Hoop and Fence Pickets	1,400 cubic feet

SHIPPED TO BRITISH PORTS

Flour	17,620 barrels
Wheat	176,268 bushels
Oats, Barley, etc.	176,268 bushels
Oatmeal	70 barrels
Gypsum, unground	324 tons
Square Timber, in vessels	8,000 cubic feet
Square Timber, in rafts	469,000 cubic feet
Flatted and round Timber	2,000 feet
Pine Lumber	2,805,000 feet
Pine Staves	31,000 No.
Pipe Staves	31,000 No.
Barrel Staves	180,000 No.
West India Staves	361,000 No.
Shingles	178 M.
Saw Logs	2,845 No.
Lath, Hoop and Fence Pickets	4,080 cubic feet
Empty Flour Barrels	4,049 No.
Cordwood	338 cords

STATEMENT OF VESSELS THAT HAVE LOADED ON THE GRAND RIVER, DURING THE SEASON OF 1850, INCLUDING SCOWS

British Vessels to British Ports		British Vessels to American Ports	
216	Tonnage 17,904	208	Tonnage 17,253
British Steamboats to British Ports		**British Steamboats to American Ports** — 55	
21	Tonnage 694		Tonnage 2,060

American Vessels

75		Tonnage 5,046
Total number of Vessels 575	Total tonnage	42,957 tons

4.11 Exports from Dunnville, 1850: Figures compiled from records at the Customs House indicate the variety and quantities of the primary products being exported through Canal ports in the Second Canal era. At this time Dunnville's population was barely 1,000, but it was obviously a busy and prosperous Feeder Canal port. Early inhabitants, eager for trade with the Americans, had loaded ships with grain, lumber, meat and cattle, and sailed off through the Great Lakes to a site on Lake Michigan, at that time unnamed. Hence lumber from the Dunnville area helped to build the future city of Chicago (incorporated 1833) — and perhaps helped to establish the link between the Niagara Peninsula and the home of the *Chicago Tribune* [see Figs. 4.8-10].
– SCHM: N-4594; R.W. Stretton, *Dunnville, Ontario: Centennial Year 1950.* Dunnville, 1950

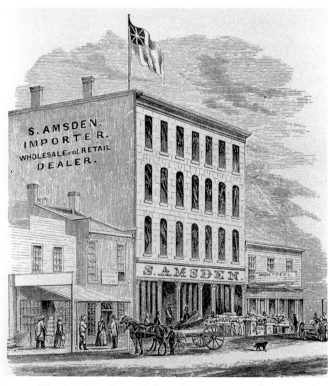

4.12 Samuel Amsden, Dunnville, 1860s: Samuel Amsden was not only a prominent importer, but also a member of the town council and customs official at this time. In the latter capacity he was responsible for compiling the table of exports in Fig. 4.11.

4.13 Port Dalhousie, ca. 1910: Moored at the Third Canal piers are "package freighters" — a development of the typical canaller, designed to fit neatly into the locks while carrying as much freight as possible. Bulk carriers and schooners cut down for use as tow barges can also be seen at this busy port which served as the northern terminus of the first three canals (1829-1932). – Burtniak Collection, *St. Catharines Journal*, 30 April 1837

4.14 Barge MAITLAND NO. 1, 1987: This cargo of beer tanks reminds us that alcoholic beverages, whisky in particular, were one of the early cargoes along the canal, along with other provisions. – SCHM: N-6239

Welland Canal Marine List.

UPWARDS.

April	Names	From	To
7	Eliza-Ann,	Port Hope,	St. Catharines.
"	Enterprise,	Oakville,	do.
"	Hamilton,	Port Hope,	do.
"	New-York,	Hamilton,	Cleveland.
9	Robert Peel,	Wellington Sq.	St. Catharines.
10	Highlander,	Port Hope,	do.
14	Gazette,	Oswego,	Huron.
"	Hudson,	do.	Cleveland.
"	Detroit,	do.	do.
"	Sir G. Arthur,	Bath,	St. Catharines.
17	O. Crook,	Oswego,	Cleveland.
"	Free Trader,	do.	do.
19	Scow,	Port Dalhousie,	Mountain.
20	John Dougall,	Toronto,	Sandwich.
21	Albany,	Oswego,	Cleveland.
"	Steph. Girard,	do.	do.
22	O. P. Starkey,	do.	do.
"	Richmond,	do.	do.
"	Elizabeth,	Toronto,	St. Catharines.
23	Eliza-Ann,	do.	do.
"	Kentucky,	Sacketts Har.	Cleveland.
"	McWhorter,	do.	Sandusky.
"	Wm. Wallace,	do.	Cleveland.
24	Caledonia,	Whitby,	do.
"	W. L. Marcy,	Sandy Creek,	do.
26	W. H. Merritt,	Port Dalhousie,	Long Point.
"	Pulaski,	Port Ontario,	Cleveland.
"	Caroline,	Hamilton,	Thorold.
"	Winnebago,	Whitby,	Cleveland.
29	Alabama,	Oswego,	do.
"	Lewis Guler,	do.	do.
"	Moses & Elias,	do.	do.
"	Aurora Borealis,	do.	do.
"	Malcolm,	do.	do.
"	Eagle,	do.	do.
"	H. Crevolin,	Sacketts Har.	Chicago.
"	Henry,	Oswego,	Detroit.

DOWNWARDS.

April	Names	From	To
"	Moses & Elias,	Gravelly Bay,	Oswego.
10	Enterprise,	St. Catharines,	Kingston.
"	Eliza-Ann,	do.	do.
16	W. H. Merritt,	Port Ryerse,	St. Catharines.
17	Sir R. Peel,	St. Catharines,	Kingston.
19	Sir G. Arthur,	do.	do.
20	Scow,	Brantford,	St. Catharines.
22	Josephine,	Cleveland,	Oswego.
24	Hamilton,	St. Catharines,	Kingston.
"	Eliza-Ann,	do.	Hamilton.
26	Hudson,	Cleveland,	Oswego.
"	Detroit,	do.	do.
"	New-York,	Huron,	do.
"	Texas,	Sandusky,	do.
27	Gen. Brock,	Grand River,	Kingston.
"	J. Woods,	do.	do.
"	Madison,	Cleveland,	French Creek.
"	Gazette,	Huron,	Oswego.
28	Malachi,	Port Burwell,	do.
"	Dawn,	Sandwich,	Kingston.

RECAPITULATION.

No. vessels passing up, in April, 1838,		37
do. down, do.		29
		66
No. vessels passing up, in April, 1837,		9
do. down, do.		7
		16
Increase, this spring,		50

St. Catharines Journal, 30 April 1837

4.15 Welland Canal, ca. 1935: Yet another staple cargo which is still carried through the canal is coal. An amateur photographer was fascinated by this coal carrier belonging to Canada Steamship Lines. Such dangerous overloading is now forbidden. – Collection of Alan Sykes

4.16 Tanker METEOR, 1972: Oil is a common 20th century cargo on the Welland Canal. From the 1890s to the 1960s "whale back" freighters, with long narrow hulls and rounded sides, were a familiar sight on the canal. While their design provided minimal resistance to wind and waves, changes in shipbuilding technology provided even better alternatives by the 1960s. The METEOR was on her last voyage when this photograph was taken. – Alfred F. Sagon-King

4.17 Thomas Rodman Merritt (1824-1906), 1873: Like his father (William Hamilton Merritt), Thomas became a miller, businessman and entrepreneur, and served as member of Parliament. He is representative of the second generation of prosperous and venturesome men whose manifold activities helped to shape the development of the Niagara Peninsula. – NAC: PA-33467

4.18-20 T.R. Merritt Interests, 1860s and 1870s: These advertisements suggest the degree of prosperity and sophistication which the Welland Canals had helped to bring about. Banking, utilities, and transportation with inter national connections: these and many other services were now based in St. Catharines, which had become a regional centre. By this time, too, business directories and gazetteers had joined the newspapers as vehicles for promotion of all manner of industrial and commercial concerns in the canal communities. – Fuller's *St. Catharines Directory for 1863-64;* Williams' *Directory for St.* Catharines . . . for 1877-78; Mitchell's *General Directory for the Town of St. Catharines . . .*, Toronto, 1865

WELLAND RAILWAY & STEAMBOAT COMPANY.

FROM
PORT COLBORNE, LAKE ERIE,
TO
PORT DALHOUSIE, LAKE ONTARIO.
☞ 25 MILES, ☜
Connecting by Steam and Sailing Vessels.
CHICAGO, MILWAUKEE, DETROIT,
And all points on the upper Lakes, with
NEW YORK, ALBANY, TROY, BOSTON, OSWEGO, OGDENSBURGH,
KINGSTON, MONTREAL AND QUEBEC,
Also connecting with the AIR LINE at Welland Junction, and Allanburgh Division, of the S. W. R. R. at Merritton.
And all other Ports on Lake Ontario and River St. Lawrence.

BOARD OF DIRECTORS.

J. W. Bosanquet, Esq., Chairman, Eng.	B. T. Bosanquet, Esq., England.
Major Kitson,	T. R. Merritt, Esq., Canada.
R. B. Wade,	Hon. Jas. R. Benson, "
Admiral Tindal.	John Brown, Esq., "

OFFICERS.
T. R. Merritt, Esq., Chairman, Local Board.
E. H. Stead, Treasurer.
Wm. Pay, Esq., Superintendent.

SECURITY
Loan & Savings Company,
ST. CATHARINES.

CAPITAL, $300,000.

DIRECTORS:

THOMAS R. MERRITT, Esq., President.
JAMES TAYLOR, Esq., Vice President.
JAMES LAMB, Esq.   ROBERT LAWRIE, Esq.
RICHARD WOODRUFF, Esq. SYLVESTER NEELON, Esq., M.P.P.
G. P. M. BALL, Esq.,
Solicitor, CALVIN BROWN, Esq.

MONEY TO LOAN ON REAL ESTATE.

Borrowers can obtain any sum they require at low rates of interest, and repay by yearly, half-yearly, quarterly, or monthly instalments.
Mortgages can be paid off at any time, according to the rules of the Company, without notice.
Full amount of Loan will be advanced, the expense of conveyancing, &c being included in the instalments if desired.
In borrowing from a Home Institution, money can be obtained with less delay than through Agents, and the risk incurred in remitting repayments is avoided.
All applications for Loans strictly confidential.

MORTGAGES PURCHASED.

ATTENTION IS ESPECIALLY INVITED TO

THE SAVINGS DEPARTMENT,

Where Interest is allowed at the rate of
SIX PER CENT. PER ANNUM.
For further information apply at the Office of the Company.
ST. PAUL STREET, ST. CATHARINES.
THOS. REYNOLDS, *Secretary-Treasurer.*

ST. CATHARINES & WELLAND CANAL
GAS LIGHT COMPANY.

James R. Benson, Esq, President.
William McGhie, Esq, Secretary and Treasurer.
Charles Millward, Esq, Superintendent.
DIRECTORS.

T. R. Merritt, Esq.	D. C. Haynes, Esq.
N. Merritt, Esq.	Dr. H. R. Goodman.
E. S. Adams, Esq.	The Mayor, (ex officio.)

The Office of the Company is on St Paul Street, nearly opposite William Street.

5.1a *Farmers' Journal and Welland Canal Intelligencer*, 22 Nov. 1836

Port Dalhousie.

THE subscriber, having laid out a *Town Plot* on an extensive scale, with regular and spacious streets, on his premises, now offers

BUILDING LOTS

For sale, to *Actual Settlers*, on moderate terms. Located on a Peninsula at the confluence of the WELLAND CANAL with Lake Ontario, with a fine commanding view of a spacious Harbour, (which will contain about 280 acres,) on one side, and the Lake on the other—a high, dry, and healthy situation—no place in the western part of America offers equal advantages to the Capitalist, the Merchant or Mechanick.

On the completion of the Welland canal, which will be the ensuing year at farthest, numerous commercial advantages will immediately present themselves. The country around is fertile and populous. Hydraulick situations, to any extent, may be erected, combining advantages no other situation on Lake Ontario can possess, as Wheat may be brought from any part thereof, deposited in the Mills, and a return load of Flour taken, with but once shifting.

Although the trade of the western countries may never centre in any given place, still this place must receive a share of it. A portion, and perhaps a greater portion of the immense wealth which will, in all human probability, be borne on the waters of the canal, will be deposited here for re-shipment across the Lake, or up the canal.

The experience of the late war, with other important considerations, have impressed every thinking man in the community, of the necessity of the canal terminating here; and accordingly the work is fast progressing under the most favourable auspices. *Dalhousie, April 5, 1826.*

9tf   NATHAN PAWLING.

---

100 Labourers
WANTED.

THE Subscribers wish to employ ONE HUNDRED good, faithful Labourers, to commence work on *Monday next*, at the Welland Canal Harbour, to whom liberal wages will be given, and prompt payment made.

HOVEY & WARD.

St. Catharines, October 28, 1826   38

TO BE SOLD, OR
EXCHANGED.

FOR Property on the Niagara Frontier, at or near the line of the Welland Canal, or at the Head of the Lake—my PROPERTY in Lobo, Delaware, and Caradoc. Enquire of Wm. H. Merritt Esq. St. Catharines.

J. MATTHEWS.

Lobo, 15th March, 1826.   81f

New Store.

THE subscriber, grateful for past favors, begs leave to inform his friends and the Publick, that he has REMOVED to his NEW STORE, opposite the "Deep Cut Hotel," where will be kept a general assortment of

Dry Goods,
Groceries, Crockery, Glass, and Hardware;
TOGETHER WITH A GOOD SUPPLY OF
WINES, BRANDY, GIN,

And other liquors—which he offers for sale, at the most *Reduced Prices*, for cash or country produce.

CARLETON H. LEONARD.

Deep Cut, Feb. 16, 1826.   21f

WELLAND CANAL HOTEL.
WILLIAM JACKES.

RESPECTFULLY informs the Publick that he has taken that well known Tavern stand, formerly occupied by the late Paul Shipman, in this Village—where he will keep a constant supply of the best of Liquors and Provisions, with other accommodations necessary for the comfort and convenience of Travellers—also, good stabling, horse-feed, &c.; and he hopes, by strict attention, to merit and receive a liberal share of patronage.

St. Catharines, February 1, 1826.   1tf

The Cottage,
[DEEP CUT—WELLAND CANAL]

J. VANARNAM, proprietor of this Establishment, thankful for past favours, wishes to inform his friends and the publick, that he continues to occupy his old stand, (near Mr. Hoard's store,) which he has recently enlarged, and improved in such a manner as to make it as comfortable and convenient for Travellers, &c. as any other in the place.

☞ Parties of pleasure, from the Falls, or any other part of the country, and other genteel company, will receive every attention the circumstances will admit, to render their stay agreeable and pleasant.

The Cottage, May 1, 1826.   13tf

5.1c *Farmers' Journal and Welland Canal Intelligencer,* - May 24, 1826

5.1b *Farmers' Journal and Welland Canal Intelligencer*, 10 May 1826

Chapter 5: Advertising and Promotion

From the mid-1820s, many industrial and business concerns developed in, or moved to, the canal-side towns to take advantage of the new mill-sites and improved communications. They were proud to announce their new situations in the early newspapers, and in the directories and gazetteers which became popular in the mid-19th century. Before street numbering became common, many firms advertised their location in relation to the canal, or to a specific lock or bridge. The communities themselves promoted their advantages, such as access to the canal for its water power or for transportation. With the advent of scenic postcards, communities and firms had themselves portrayed in hundreds of views (many featuring the canal and its locks) which, as they well knew, would be sent far and wide. The federal Department of Railways and Canals also permitted hundreds of pictures of the canal to be reproduced on post cards: the construction of the Fourth Canal became a favourite theme in the 1920s and 1930s. With increasing use of the telephone, the popularity of post cards has declined, although even today communities and the Welland Canal, as part of the St. Lawrence Seaway, are featured on glossy full-colour cards.

5.1 Newspapers, 1820s and 1830s: In the First Canal era, aside from broadsides, and posters at specific locations (both of which had limited audiences), newspapers were the main vehicle for advertising and promotion. Plans for the proposed canal — "Mr. Merritt's Ditch" — were described and debated in various newspapers throughout the Niagara Peninsula [see map, Chap. 2, Fig. 2.6]. For example, an editorial in the Niagara *Gleaner* of 14 August 1824 began as follows:

We are informed the whole intended route of the Canal is now surveyed. From the Grand River to Chippawa is ten miles, along the Chippawa to Lake Ontario by the Twelve Mile Creek, or to the river at this Town [Niagara], twenty miles.

There is but little difference in the length from the Chippawa to the mouth of the Twelve Mile Creek or to the River at this Town [Niagara].

There is also but little difference in estimated expense of the route to those two places. That to this Town [Niagara], we understand is estimated at about £2,000, more than that to the mouth of the Twelve Mile Creek; in the last route, however, there is a much greater part to be done on perishable materials than the former, besides the greater superiority of the noble harbour, formed by nature in the River, and the superficial one to be made at the mouth of the Twelve Mile Creek.

This will, we have no doubt, induce the Directors to bring it to Niagara.

5.1d *Farmers' Journal and Welland Canal Intelligencer,* 22 Nov. 1826

On February 1, 1826, St. Catharines' printer and publisher Hiram Leavenworth produced the first issue of the *Farmers' Journal and Welland Canal Intelligencer* — William Hamilton Merritt's "mouthpiece" for his canal [Fig. 5.1a]. It and other Peninsula newspapers discussed the canal itself, its progress (or lack thereof), and the advantages it would bring to the area. Real estate sales [Fig. 5.1b] and the development of communities along the route of the canal were promoted through the papers. The canal received additional publicity from many individuals who, in advertising their assorted goods and services [Figs. 5.1c and d], touted their location at or near the canal.

Public Notice

IS hereby given, that all TOLLS must hereafter be paid in Specie, or in Notes of the Banks of Upper and Lower Canada, or of the Welland Canal Company. By order of the Board.

P. G. BEATON, *Secretary.*

Welland Canal Office,
St. Catharines, 8th Sept., 1837.

☞ The Editors of the Commercial Herald, *Oswego*, N. Y.; the Herald, *Cleveland*, Ohio; the Courier, *Detroit*, Michigan; the Patriot and Correspondent, *Toronto*, U. C. and Reporter, *Niagara*, will please give the above notice *four* insertions in their respective papers, and send one copy containing the first insertion, with their bills, to the Canal office for payment.

5.1e *St. Catharines Journal,* 28 Sept. 1837

Welland Canal now open!

THE publick is respectfully informed, that the Welland Canal is now *free from ice*, throughout its whole extent; as also the harbours, at each extremity; and that the water having been kept at full head, and no extensive repairs required, during the winter past, this channel is now in complete readiness for immediate navigation.

GEO. PRESCOTT, *Sec'y.*

Welland Canal Office,
St. Catharines, April 2, 1838.

The following papers will insert the above notice, *three* times, and send their accounts to the canal office : *Detroit* Morning Post and Com. Herald, *Cleveland* Herald, *Buffalo* Daily Advertiser, *Oswego* Palladium, *Toronto* Patriot, *Kingston* Chronicle, *Brockville* Statesman, and *Prescott* Sentinel.

5.1f *St. Catharines Journal, 26 April 1838*

At Port Colborne.

A *STONE GRIST MILL*, with two pairs of French Burr Stones, and all the necessary Machinery for Country and Merchant work—cast iron gearing throughout. The Mill is 30 by 40 feet, and three stories high, with an *addition* of 55 by 30 feet, for a STORE-HOUSE. Also, a stone built ENGINE HOUSE, two stories high, 24 by 40.

CARDING, FULLING & CLOTH DRESSING WORKS, complete.

A *SAW MILL*, complete, and capable of cutting 6,000 feet in 24 hours.

These Mills and other Machinery are driven, during the summer months, by a *STEAM ENGINE* of 20 horse power, which is a first rate one, and capable of working up to a much greater power, if a new boiler were added. During the Winter, and high water in the Spring and Fall, they are driven by the waste water of the canal.

A number of *VILLAGE LOTS*, also, for sale, at this place, *Port Robinson* and *Marshville*—together with several valuable Lots of Land, containing 50 acres each, lying on the canal and feeder.

Tenders for Leasing the Grist and Saw-Mills, together or separately, also the Clothing and Carding Works, will be received at this office up to the *1st October* next—where also the price of Village and other Lots will be made known.

For further particulars apply either personally or by letter, to J. G. STOCKLY, at *Allanburgh*—JAMES BLACK, Esq. at *Port Colborne*—JOHN CALLAGHAN, *Port Robinson*—JOHN GRAYBIEL, *Marshville*—JOHN CLARK, Esq. *Port Dalhousie*—or at this office, to P. G. BEATON, *Secretary.*

Welland Canal Office,
St. Catharines, 5th Aug. 1837.

5.1g *St. Catharines Journal,* 8 March 1838

By the late 1830s, with the Welland Canal in operation, newpapers such as the *St. Catharines Journal*, while still carrying notices regarding the Canal itself [Figs. 5.1e, f, g], and real estate [Figs. 5.1h and i], could feature a much broader and more sophisticated range of goods [Fig. 5.1j]. The Niagara Peninsula was being opened to a wider world [Fig. 5.1k].

ALLANBURGH FURNACE.
WELLAND CANAL, U. C.

THE Furnace business heretofore carried on by THOMAS TOWERS, will, from the 1st July, inst. be conducted by the Subscribers, under the firm of "THOMAS TOWERS & Co."

THOMAS TOWERS,
JOHN L. JAMES,
By his Attorney, THO's TOWERS,
WM. A. CLEMENTS.

Allanburgh, July 20, 1837.

5.1h *St. Catharines Journal*, 28 Sept. 1837

Custom House Sale.

WILL be sold, at publick Auction, at the house of the COLLECTOR, in St. Catharines, on Tuesday the *fifth of February,* 1839, at the hour of 12 o'clock, noon, the following articles, viz :

20 bags containing TEA ;
7 small chests   do.
7 paper parcels   do.
1 double WAGON, with 1 double
    Whiffletree, 2 single do. and 1 log Chain ;
1 sett double HARNESS ;
2 Buffalo ROBES ; 1 Whip ;
1 horse Rug and Circingle.

JOHN CLARK, *Collector.*
Collector's Office, St. Catharines, Jan. 2, 1839.

5.1j *St. Catharines Journal*, 17 Jan. 1839

FOR CHICAGO,
AND OTHER PORTS ON LAKE MICHIGAN.

THE fast sailing Schooner, VICTOR, Capt. F. DUBOIS, will leave this place, for the above ports, on the 10th *day of June,* 1839. This vessel is good and well found, and has good accommodations for passengers. For freight or passage, apply to L. DYER, *St. Catharines.*

5.1k *St. Catharines Journal,* 13 June 1839

VALUABLE PROPERTY,
IN DUNNVILLE,
FOR SALE, OR TO RENT.

THE Subscriber offers for Sale, or for Rent, on very advantageous terms, and possession given as may be agreed upon, the whole of his valuable Property, situated in the flourishing village of Dunnville, at the Grand river dam, head of the Feeder to the Welland canal, consisting of a strong and substantial

SAW-MILL,

40 by 60 feet, with two separate Saws, capable of turning off 50,000 feet of Lumber per week. In the same building is a SHINGLE MACHINE—CIRCULAR SAWS, for cutting up slabs into Lathing, to the best advantage ; and also, *Saws* for cutting out all sorts of Carriage stuff; and a *Lathe* capable of turning timber from two to eighteen feet in length.

Along side the Mill, vessels of any burthen may lay to and take in cargo ; and from its contiguity to the Grand river and Feeder of the canal, in addition to its site having great water privileges, and the opportunity of obtaining an abundant supply of logs, delivered at the Mill, from the upper part of the Grand river country, this Establishment deserves the attention of any capitalist who wishes to make a profitable investment. In connexion with this property, there is also a *Log Harbour,* well boomed and piled, situated about a quarter of a mile up the river, sufficient to admit 20,000 Saw logs.

A STORE-HOUSE, with a MERCHANTS' SHOP in front, at the junction of the river and feeder, 36 by 46 feet square, now occupied by *Dittrick & Chisholm,* which, together with a commodious building 24 by 36 feet, formerly occupied by the Messrs. Clarkes, on the adjoining lot and wharf, as a Merchants' Shop, afford a most excellent chance of doing good business.

5.1i *St. Catharines Journal*, 28 Sept. 1837

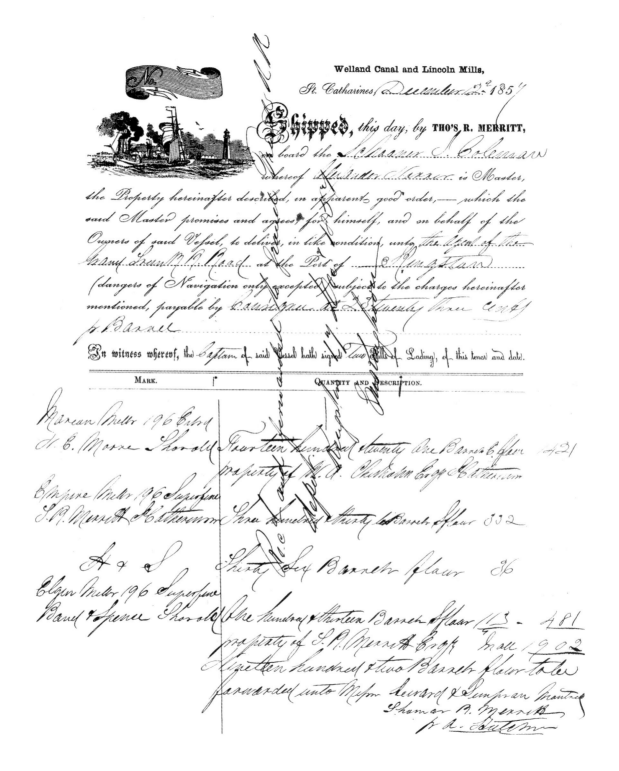

Welland Canal and Lincoln Mills,

St. Catharines, *December 2d 1857*

Shipped, this day, by **THO'S R. MERRITT,** on board the *Schooner J. Coleman* whereof *Alexander Nazzar* is Master, the Property hereinafter described, in apparent good order,—— which the said Master promises and agrees for himself, and on behalf of the Owners of said Vessel, to deliver, in like condition, unto *the Mash of the Grand Trunk R.R. Road* at the Port of *Kingston* (dangers of Navigation only excepted) subject to the charges hereinafter mentioned, payable by *Consignee at Twenty Three Cents pr Barrel*

In witness whereof, the Captain of said Vessel hath signed Two Bills of Lading, of this tenor and date.

MARK.	QUANTITY AND DESCRIPTION.	
Marian Mills 196 Extra N.E. Morse Thorold	Fourteen hundred twenty One Barrels Flour property of W. A. Chisholm Esqr Chatham	421
Empire Mills 196 Superfine J.R. Merritt St Catharines	Three hundred thirty two Barrels Flour	332
A & S	Thirty Six Barrels flour	36
Elgin Mills 196 Superfine Ranel & Spence Thorold	One hundred thirteen Barrels Flour property of J.R. Merritt Esqr	113 — 481

In all 1902
Eighteen hundred & two Barrels flour to be forwarded unto Messrs Heward & Simpson Montreal

Thomas R. Merritt
W. A. Eaton

5.2.b SCHM: Norris Papers

66

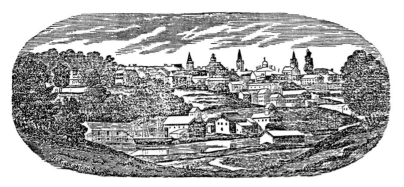

5.2 Newspapers and Directories, 1850s: By the Second Canal era, the newspapers commanded a much wider audience, augmented in part by the growing communities along the canal route. In some cases, newpaper mastheads featured canal views, either imaginary or, as in the case of the *St. Catharines Constitutional*, realistic [Fig. 5.2a, and see also Chap. 6, Fig. 6.1]. The growing audience was increasingly aware of that wider world beyond the Peninsula: through the papers they read, of course, but also through the expanding variety and sophistication of goods and services available by means of the canal, and through the greater ease of travel, which brought visitors from near and far.

5.2a *St. Catharines Constitutional*, 4 Oct. 1866

Partnerships became more common in business and industry, and the use of printed business forms, often featuring appropriate (and often elaborate) decoration, became popular with banks, shipping companies, mills [Fig. 5.2b], and so on [Fig. 5.2c]. In addition, business and trade directories — more permanent and convenient than the fragile and often cumbersome newspapers — entered the promotional field. Whether the directories were local, county, provincial or national, they had the added advantage of including information on other areas and useful general information on such subjects as banks, post offices, coach service and insurance agents.

5.2c SCHM: N-2798

Again, location on or near the canal, as well as the advantages of the canal in both importing and exporting, were proclaimed [Figs. 5.2d, e, f, g].

TO SHIP OWNERS
AND
MASTERS OF VESSELS.

JAMES McCOURT,
SAIL MAKER AND RIGGER,
Welland Canal, near St. Paul Street,
ST. CATHARINES.

Would respectfully intimate that he has established himself in the above business, at St. Catharines, C.W., where he is prepared to attend to all orders in his line, with punctuality, durability and dispatch. J. McC. having had over 25 years' experience in the business, both in Canada and the United States, and by keeping the best materials on hand, feels confident of giving general satisfaction.

N.B.—All materials furnished if required, and all work warranted. On hand, at all times, a choice assortment of Tarred and Manilla Ropes, Ships' Blocks and Canvass from No. 1 to No. 10, Sails, Twine, Needles, &c., &c.

5.2e Evans' *Gazetteer and Business Directory* of Lincoln and Welland Counties, 1879

HENRY MUSSEN
RETAIL DEALER IN
GROCERIES & PROVISIONS,
CONFECTIONERY, GENUINE PATENT MEDICINES, CUTLERY,
GARDEN SEEDS & IMPLEMENTS,
EARTHENWARE,

And all articles usually found in a General Store, all of which will be sold as Cheap as any House in the County of Welland.—Terms Cash.
CLOCKS AND JEWELRY NEATLY CLEANED & REPAIRED.

☞ REMEMBER THE PLACE: ☜
WEST SIDE OF GUARD LOCK, : : : ALLANBURGH, ONT.

5.2f Horwitz, *St. Catharines General and* Business Directory, 1874

St. Catharines Oil Refinery.
H. F. LEAVENWORTH, Proprietor,
WHOLESALE DEALER IN
Canadian, American & Signal
Illuminating Oils, Machine Oil,
NAPHTHA, &c.,
Office and Refinery, Opposite Lock No. 2, Welland Canal.
ST. CATHARINES, ONTARIO.

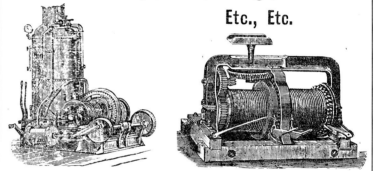

ROBERTSON BROS.
WELLAND, ONT.
MANUFACTURERS OF
Steam Hoisting Engines, Portable Engines
Horse Power Hoists, Hand Hoists
Derricks, Derrick Fittings,
Etc., Etc.

Various sizes and styles for use by Contractors, Railroads, Builders, Stone Dealers, Quarrymen, etc., for handling material in building and excavating, or on steam ships, barges, scows, docks, warehouses, coal yards, or any purpose where power is required for hoisting and handling material.

ESTIMATES FURNISHED

For special arrangements for special purposes, and for gangs of one or more drums for operating several derricks from one power by belts, gearing or coupling. **Correspondence Solicited**

ALSO
FURNACES
(COAL OR WOOD)

For heating dwellings, public buildings, etc. Low down, all cast iron, simple, substantial, and thoroughly original, Easy to set up and a good seller. Constructed on scientific principles of correct material, giving thorough utilization of fuel and greatest economy. **Write for Particulars**

5.2d Fuller's *St. Catharines Directory for 1863-64*

5.2g *Ontario Gazetteer, 1892-93*

5.3 Late 19th-Early 20th Century: In this period newspapers, trade and business directories, and magazines were joined by promotional "bird's-eye" maps, ringed with representations of prominent civic and commercial buildings [Fig. 5.3a; see also Introduction, Fig. Int. 2; Chap. 3, Figs. 3.17-19; and Chap. 11, Fig. 11.16 for other drawings from this map]. Historical atlases also appeared: that for Lincoln and Welland pictures a number of canal-side establishments; that for Haldimand and Norfolk, includes business cards, a number of which give their location near the canal [Fig. 5.3b]. Such publications would find their place on library and Mechanics' Institute shelves, along with the more traditional forms of public information, further promoting the canal, its communities, and their trade and commerce.

5.3a St. Catharines Centennial Library

5.3b Page's *Illustrated Historical Atlas of the Counties of Haldimand and Norfolk*, 1877

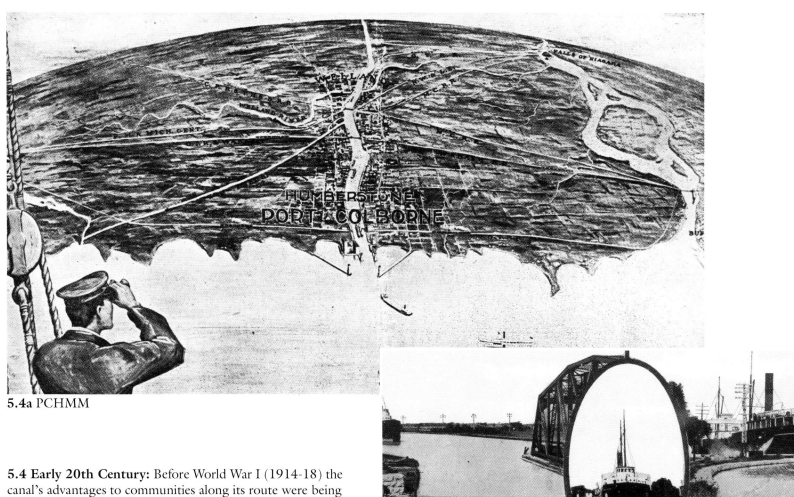

5.4a PCHMM

5.4 Early 20th Century: Before World War I (1914-18) the canal's advantages to communities along its route were being promoted in various media [Fig. 5.4a]. By now commercial reproduction of photographs, rather than line drawings or etchings, had greatly increased the range, variety and sophistication of illustrations.

5.4.b Burtniak Collection

Postcards, especially the scenic variety, were increasingly popular, and canal communities and their business and industrial concerns were quick to tout the advantages of their favourable locations [Figs. 5.4b, c, d, e]. With construction of the Fourth Canal (beginning 1913) the government allowed reproduction of official construction photographs, to publicize the Ship Canal [Fig. 5.4e].

Port Colborne Harbor

3536

SWING BRIDGE, THOROLD, ONT.

Since ordinary people made frequent use of postcards to send a tremendous variety of messages far and wide, a growing audience was made aware of the advantages to be found in the Peninsula. The view of the NORONIC [see Chap. 9, Fig. 9.12], for example, was bought in Port Colborne, posted in Sault Ste. Marie, and sent to Zurich, near the eastern shore of Lake Huron.

5.4c, d, e Burtniak Collection

W. C. 14. Welland Ship Canal. Lock No. 8, looking S. from Main St., Humberstone.

5.5 The Later 20th Century: As hydro-electricity began to replace steam as a source of power, Niagara Peninsula civic and business interests were quick to advertise one of their greatest advantages: their proximity to the cheap "hydro"first generated at Niagara Falls in the early years of the 20th century [Fig. 5.5a]. The availability of the new power source did not, however, mean the end of the canals' attractions. Many industries both long-established and new, continued to make use of the water supplied by the old canals, for washing and cooling, as well as for removing effluent. And of course the Fourth Canal continued its traditional role in transportation [Fig. 5.5b]. In recent years the older media have been augmented by film and videotapes, produced by various levels of government, other public bodies, and private firms. The Welland Canal, part of the St. Lawrence Seaway since its opening in 1959, and the communities and businesses along its route, have been seen in both old and new media, reaching an ever-wider audience, across North America and abroad.

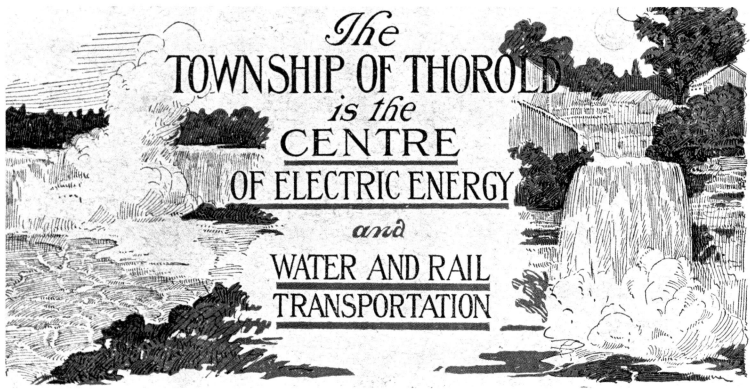

5.5a AO: Pamphlet 1920, no. 76

5.6 Postage Stamps: (opposite) The role of the Welland Canal as a crucial part of the St. Lawrence Seaway was recognized by Canada Post on the occasion of the Seaway opening in 1959 [Fig. 5.6a] and again during the 25th anniversary of that opening, in 1984 [Fig. 5.6b]. In 1974 a stamp was issued to commemorate the 150th anniversary of the start of the First Welland Canal and William Hamilton Merritt's importance in its realization.

Throughout the Welland Canal's history, local and regional individuals and groups, and the Government of Canada, have, in the process of promoting their own interests, helped to spread awareness of the waterway's importance to the communities of the Niagara Peninsula, to the country as a whole, and to other nations of the world, using the whole range of media available.

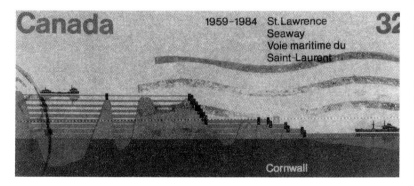

5.5b SCHM: Welland Canal Archives

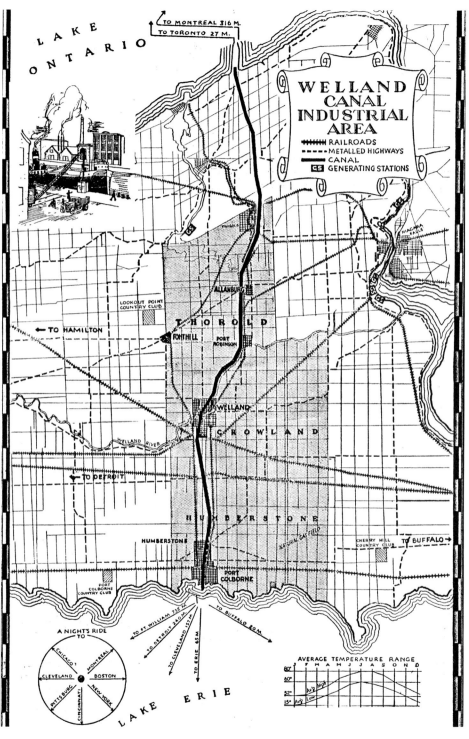

Chapter 6: Communities, Initial Stimulus

6.1 St. Catharines, 1853: A new line of settlements began to develop at the main construction sites along the the First Welland Canal. The only settlement on the canal route which predated construction of the waterway was a sleepy Loyalist village where the main east-west road crossed Twelve Mile Creek, "Shipman's Corners." The Welland Canal stimulated rapid growth at this crossing-point, growth which has been sustained to the present. The raceway in the lower right corner is part of the system which fed mills such as the Merchants' Mill on the right [see Chap. 4, Fig. 4.1 for a later structure on the same site]. The mills in turn were dependent on the trade of the lake schooners which passed through the canal. – *Anglo-American Magazine*, August 18, 1853

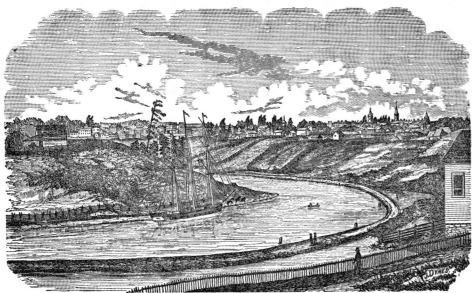

VIEW OF ST. CATHERINES, C. W.

6.2 St. Catharines, 1871: Twenty years later the boom continued as shops and businesses of brick and stone lined St. Paul Street (formerly an Indian trail). The offices of the Quebec Bank (founded in 1818) suggest the sophisticated economic life which the canal was bringing. Many of the buildings on the right backed on to the canal and its wharves. – *Canadian Illustrated News*, 23 September 1871

WOOD BROS.,

DEALERS IN

CHOICE FAMILY GROCERIES

AND PROVISIONS.

China, Crockery and Glassware.

ORDERS FOR

SHIP STORES AND MESSES,

when in season, promptly attended to on the most reasonable terms.

YOUNG'S OLD STAND,

FRONT STREET, - PORT DALHOUSIE.

6.3 Port Dalhousie, ca. 1870: When the mouth of Twelve Mile Creek became the northern terminus for the First Canal, a thriving village sprang up, serving and dependent on canal traffic. Several businesses, including Wood Brothers (centre background) and their on-site successor, Murphy's, specialized in ship provisions. A number of hotels and taverns catered to the crews, and shipyards built and repaired their vessels. – NAC: PA-14526; Williams, *Directory of St. Catharines . . . for 1877-78*

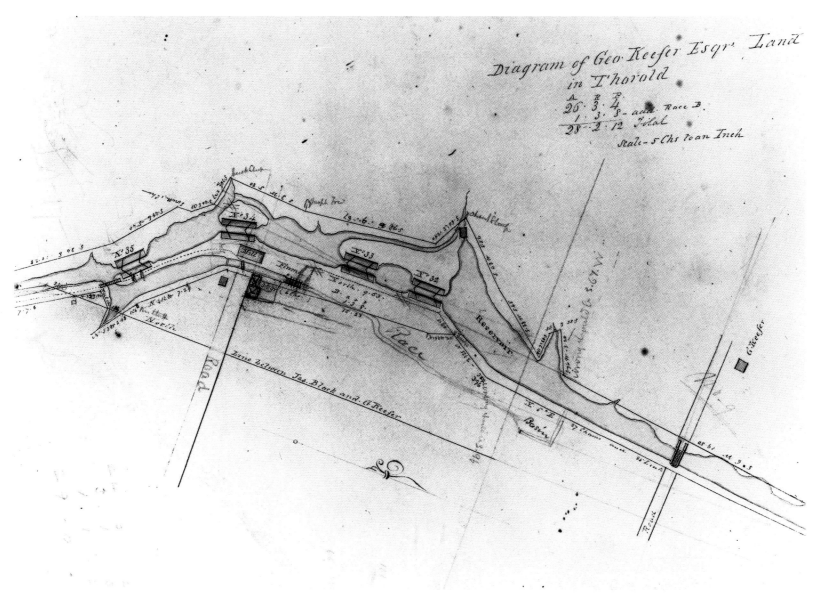

6.4 Thorold, c. 1830: In 1827 George Keefer built his flour mill at the future site of Lock 34, even before the First Canal came through. Here the mill and its associated raceway and flume mark the heart of the later village. The bridge taking St. David's Road over the canal was one of the few in the area, and hence became a hub of business. A later Keefer house (Maplehurst) still stands on the site indicated here for George Keefer's home. – *Survey of Lands Appropriated to the Use of the Welland Canal Company, 1826*

6.5 Thorold: This community was created by the canal, it was supplied by the canal, and in turn it supplied canal traffic. By the 1870s many Thorold businesses advertised to commercial interests, stressing their utility to shippers, as well as the convenience of their canalside locations. – Horwitz, *St. Catharines General and Business Directory for 1874*; Evans' *Gazetteer and Business Directory for Lincoln and Welland Counties*, 1879

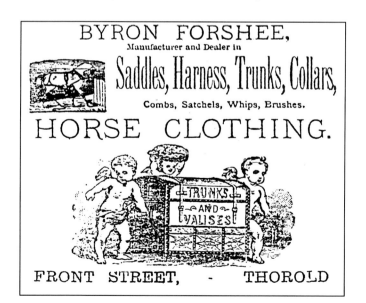

6.6 Black Horse Corners, 1928: This tavern, at a crossroads east of what became Allanburgh, hosted the banquet following the turning of the first sod for the Welland Canal, on 30 November 1824. The lengthy speeches at the sod-turning are quoted in Merritt's *Biography* [p. 67], which continued: "After three cheers, the company adjourned to the Inn, where a very good dinner was served by Mr. Beadgerley, to thirty-four gentlemen. George Keefer, Esq., and John Clark, Esq. [MPP], did the honors of the table. After the cloth was removed, toasts were proposed and unanimously carried, when the company separated about dusk, highly pleased with the transaction of the day." The tavern has long since been demolished, its place in the history of the Welland Canal forgotten. – SCHM: N-7272

6.7 Port Robinson, July 1920: For nearly a century, Port Robinson (at the junction of the canal and the Welland River) was a focus of shipbuilding and up to 1897 a port of entry. Until 1833, when the waterway was extended to Port Colborne, it was the southern terminus of the canal. Its several hotels continued to provide a welcome break here at the halfway point between the lakes, for the towboys and their horses, and the ships' crews and passengers. Here, dredges, a freighter, and tugs suggest a lively trade and commerce, today vanished. – *Jubilee History of Thorold*

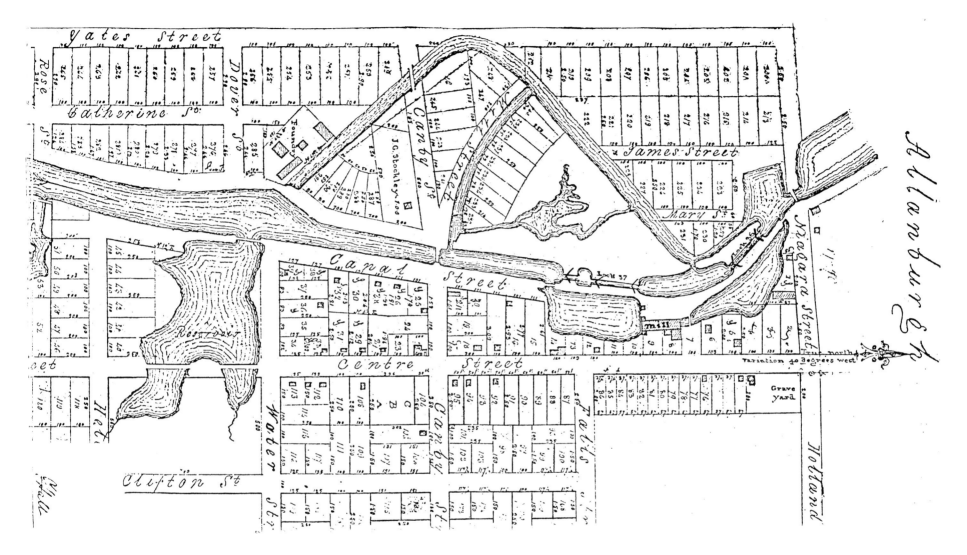

6.8 Allanburg, ca. 1836: Surveyed by engineer Samuel Keefer and drawn by Francis Hall, this map of Allanburg's numbered lots suggest plans for a large, bustling community. Allanburg was one of the settlements created by the canal, but it prospered only briefly and then declined [see Chap. 7, Fig. 7.10]. The village supported a foundry and mills, but never developed to the extent that the map indicates. Note that one of the proposed streets was named after the American financier J.B. Yates [see Chap. 2, Fig. 2.9]. – *Survey of Lands Appropriated to the Use of the Welland Canal Company, 1826*

Welland, East side Canal.

6.9 Welland, ca. 1910: The small settlement of "Seven Mile Stake" became "Aqueduct" when the Feeder Canal crossed the Welland River in 1829 and continued to grow. Known as Merrittsville after 1842, it became the county seat for Welland in 1855 and received its present name in 1858. Growth was continuous until the 1930s Depression and included construction of handsome public and commercial buildings along Main Street and, in the foreground, Canal (later King) Street. The structure with the steeple is City Hall (1898-1959). The Third Canal passes over the Second and Third Canal aqueducts (at left), and under the Alexandra Bridge (foreground). – Burtniak Collection

6.10 and 11 Chippawa, ca. 1910 and 1873 (opposite)**:** Traffic from the Niagara River could reach the Welland Canal through the Welland River. The village of Chippawa, at the confluence of the two rivers, was an important port for the canal until 1833 [see also Chap. 1, Fig. 1.13]. It supported tanneries and a distillery then, and served the canal traffic embarking up the Niagara to Buffalo. Since this brief heyday, it has remained a recreation centre, now part of Niagara Falls. We can only speculate as to the town's fate, had public pressure succeeded in 1873. – Burtniak Collection

View showing Part of Bridge, Chippawa, Ontario.

CANAL ENLARGEMENT!

At the request of the People a Public

MEETING

Has been called by the Mayor, and will be held at the

Town Hall, Clifton, Ont,
FRIDAY, JANUARY 17, 1873.

To consider the subject of Canal enlargement, and to adopt measures to lay before the Government the advantages of the route from

CHIPPEWA TO QUEENSTON.

There can be no doubt this is the SHORTEST, CHEAPEST & BEST ROUTE for a Canal connecting the two Lakes Erie & Ontario. The present opportunity may never occur again, and therefore requires the

United, Earnest and Immediate Action
of the People of the Niagara Frontier.

The People of Niagara, Queenston, Clifton, Stamford, Chippewa, Willoughby, Bertie and Fort Erie, are especially invited to Attend the Meeting, and the public generally are invited.

Turn Out En-Masse!

And let the Government and Country see that you are in earnest in this great and important matter.

CHAIR TO BE TAKEN AT 7·30 P. M.

6.12 The Feeder Canal: From 1829 to 1887 the Feeder provided the Welland Canal with water from the Grand River. In the process, it supplied water power for mills and carried ships between the Grand River and the Welland Canal. From the junction of the Feeder and the Welland River, traffic could proceed along the Welland River to the Niagara River, and on to Buffalo. – Loris Gasparotto, Department of Geography, Brock University

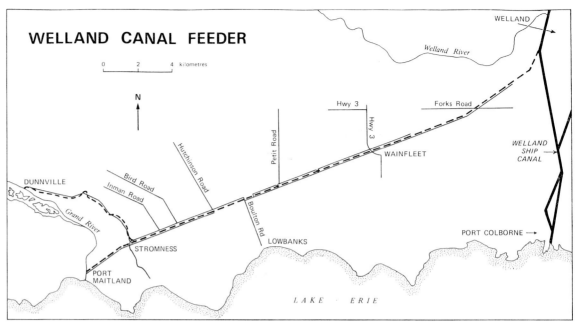

WELLAND CANAL FEEDER

0   2   4   kilometres

N

WELLAND

Welland River

Hwy 3

Forks Road

Petit Road

Hwy 3

WAINFLEET

WELLAND SHIP CANAL

DUNNVILLE

Bird Road

Inman Road

Hutchinson Road

Grand River

Boulton Rd

PORT COLBORNE →

STROMNESS

LOWBANKS

PORT MAITLAND

L A K E   E R I E

6.13 Stromness, 1989: A small village prospered for a time where the Dunnville and Port Maitland branches of the Feeder converge in Haldimand County, but today it has nearly disappeared. – R.M. Styran

6.15 Wainfleet, 1989: In 1880 a modestly assertive town hall was built where the road meets the Feeder. By this time the Feeder was in decline, but the village remained an agrarian centre. [See Chap. 7, Fig. 7.14 for another contemporary view.] – R.M. Styran

6.16 Wainfleet, ca. 1910: Entrepreneurs were drawn to Marsh-ville, where the Feeder met a road and also provided water power for mills. A then modern bridge was a landmark in the area. It would also appear to have been a source of community pride, or at least of some interest to a local teacher, who poses here with her school class. – Burtniak Collection

6.14 Stromness, 1989 (opposite): In 1842 Lauchlin McCallum emigrated from Scotland and worked as a construction contractor on the Second Canal. In 1855 he settled at Stromness, where he operated a sawmill for many years and built his home. McCallum was also involved in shipbuilding (his sawmill produced ship timber, among other products). He organized, and for a time commanded, a volunteer naval company which was involved in the skirmishes with Fenian raiders at Fort Erie in 1866. Over the years McCallum was active in local politics and served as both MPP amd MP. Today this elegant house is remote and no recognition is given to a once prominent individual. – R.R. Taylor

6.17 Port Maitland, ca.1910: Since a British naval base stood at the mouth of the Grand River from 1815 to 1834 , the Welland Canal Company could not use this site as a terminus for the Feeder Canal. After the naval authorities relinquished control, the

6.18 Dunnville, 1912: The three weir ponds and their mills were still in use, with the Feeder guard lock to the left and the dam to the right (out of sight). The flourishing town is behind the cameraman. – Burtniak Collection

6.19 Dunnville, 1834: The Grand River Dam, which channeled water into the Feeder Canal, also provided power for mills which became the heart of a new community, acting as its commercial and social centres. By 1879 similar development had occurred at the south end of the Dam [see Introduction, Fig. Int. 15]. – *Survey of Lands Appropriated to the Use of the Welland Canal Company, 1826*

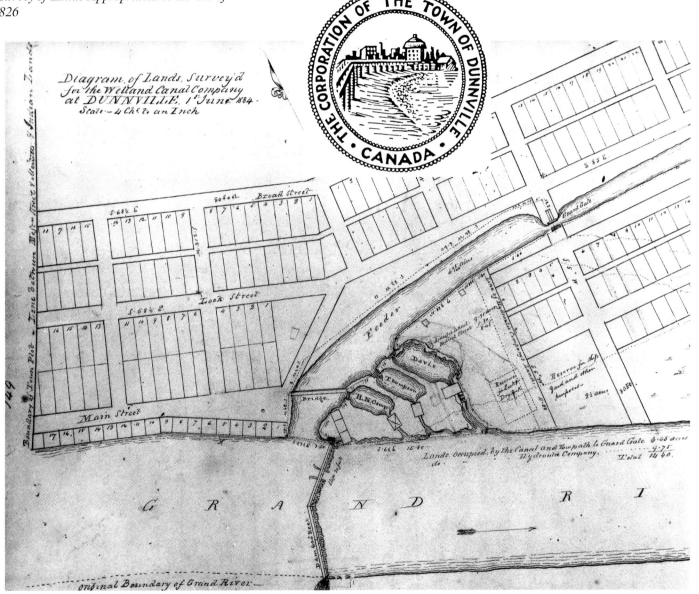

MAIN ST. E.   HUMBERSTON ONT

6.20 Humberstone, ca. 1915: Known as Stonebridge until 1851, when it became a postal village, Humberstone was absorbed into Port Colborne in 1952 as the southern terminus of the canal expanded. – Burtniak Collection

6.21 Humberstone, 1879 and 1888: Humberstone businessmen, like their fellows in other canal-side towns, took advantage of the various directories to advertise their goods and services. – Evans, *Gazetteer and Business Directory of Lincoln and Welland Counties . . . for 1879;* Morrey's *Directory for the Counties of Brant, Haldimand, Halton, Lincoln, Welland and Wentworth, 1888*

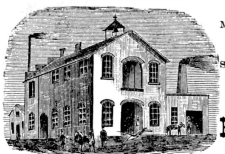

6.22 Port Colborne, ca. 1865: At the southern terminus of the canals, a number of businesses developed near the Second Canal-era swing bridge. They included a general store and later a hotel. – PCHMM: 980.4

Chapter 7: Communities, Changing Fortunes

7.1 Port Dalhousie, ca. 1890: As the waterway underwent successive enlargements and changes of route, the communities it had spawned experienced varying fortunes. Here, at the northern terminus of the first three canals, a dredge, a tug and the GARDEN CITY have put in at Muir Brothers' Dry Dock, which was founded in 1850 and operated until 1954. The town supported a variety of industries, including two shipyards and a flour mill. After the opening of the present canal and the cessation of the Toronto steamer traffic in 1958, however, "Port" declined, until its rebirth as a tourist resort in the late 1970s. [On the new GARDEN CITY, see Chap. 9, Fig. 9.13] – SCHM: N-4856

7.2 Port Dalhousie, 1989: The renovated and expanded Port Mansion Hotel, formerly the Port Hotel, symbolizes the town's rebirth. The carnival attractions of Lakeside Park have vanished – save for the carousel [see Chap. 9, Fig. 9.9] – but a thriving marina and the old canal piers attract locals and outsiders alike. – R.R. Taylor

7.3 St. Catharines, 1989: Promotional maps are far from new [see Chap. 5, Fig. 5.3], and a contemporary full-colour poster typifies concerns in "The Garden City" today: services (including "fast foods"), communications, recreation, ethnic diversity, and heritage. This top right detail of the 11.02 by 14.96 cm (28 by 38 in) poster shows expansion into previously rural land. The "old" city, which spread out from the vicinity of the first two canals, takes up most of the left-hand area of the complete map. While the city's industries have long since outgrown dependence on the Welland Canal, the impetus it provided has helped St. Catharines become a major regional centre. The St. Lawrence Seaway Authority is a major employer, with several hundred people on its payroll. – An Artistic Characterization of the History and Features of the St. Catharines Area, by Descartes Inc., Grimsby, Ontario

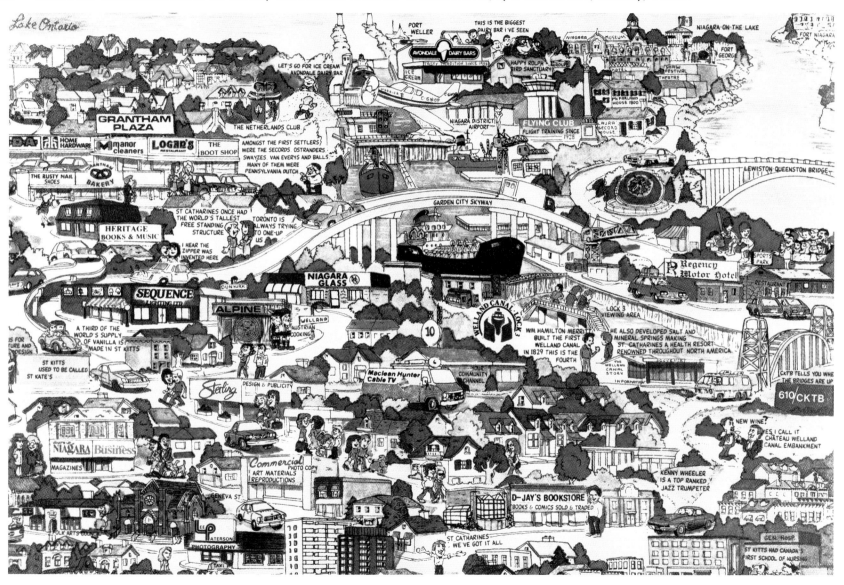

BEAVER MILLS,

MERRITTON, Ont.

W. W. WAITE, Proprietor.

MANUFACTURER OF

YARNS, BATTING AND WADDING.

7.5 Merritton, 1874: Beaver Mills, established 1857, was one of two cotton mills in Merritton in the latter part of the 19th century. The proprietor became the first reeve of the village when it was incorporated in 1874. – Horwitz, *St. Catharines General and Business Directory for 1874*

7.4 Merritton, ca. 1860: In the 1850s, a thriving industrial corridor began to develop, with the fifteen Second Canal locks in what became Merritton providing essential water power for mills. The sawmill of Noah and O.J. Phelps [see also Chap. 3, Fig. 3.20], established in 1855, was a typical example of the way in which pioneer industry was galvanized by the waterway. The Phelps brothers, descendants of a First Canal contractor, were entrepreneurs who helped to create this industrial corridor. – SCHM: N-1034

7.6 Merritton, ca. 1910: The stately town hall (1879) and the hotel with its welcoming porch were the results of a boom created by the industries along the Second Welland Canal. In the distance, at the curve of Merritt Street, stands the Beaver Cotton Mill, one of a score of enterprises supported by canal water power and trade. The town's commercial viability endured well past the construction of the Third and Fourth canals, both of which by-passed the town. Both Merritton and Port Dalhousie became part of the city of St. Catharines in 1961. – Burtniak Collection

MERRITT ST., MERRITTON, ONT.

7.7 Thorold, ca. 1925: The Third Canal had by-passed the town after 1887, and the Fourth Canal was now under construction. The canal-side mills were in decline (ships having ceased to pass through about 1915), and the waterway would soon be covered over. The town, however, continued to prosper as a paper-milling centre. – SCHM: N-5285

7.8 Thorold, ca. 1930: Despite the Depression, Thorold's Front Street, which turned its back on the old waterfront, had a settled and prosperous look. The trestle crossing the road in the distance carried the Niagara, St. Catharines & Toronto Electric Railway through the heart of the former "Stumptown." After about 1900 cheap hydro-electricity from Niagara Falls replaced water as a source of power, but railway and later highway links with eastern Canada and the United States have kept Thorold bustling. As with Welland, removal of the waterway from the town centre has not meant continued decline. – Koudys Collection

7.9 Thorold, 1989: Buoyed at first by lumber and flour milling, Thorold's economy has, in the 20th century, become devoted to papermaking. The Fraser factory (established in 1912 as Provincial Paper), on the site of several former paper mills, is a good example of this development. Fraser, now owned by Noranda Forest, is currently in the vanguard of paper recycling. – R.R. Taylor

THE OLD WELLAND CANAL
LE VIEUX CANAL DE WELLAND

Originally conceived in 1818 by its promoter, William Hamilton Merritt, to divert trade from the Erie Canal and New York and built under private auspices, the canal was opened to traffic in 1829. After additional work in 1833, the canal with its 40 wooden locks linked Port Colborne on Lake Erie and Port Dalhousie on Lake Ontario and brought prosperity to its environs by permitting the export of Upper Canadian staples through New York. In 1841 reconstruction was begun by the Canadian government to improve the canal's military and commercial value.

Dès 1818, William Hamilton Merritt conçut l'idée de ce canal pour détourner le commerce du canal Érié et de New York. Des intérêts privés en assurèrent la réalisation et on l'ouvrit en 1829. De nouveaux travaux, en 1833, permirent de relier Port Colborne, sur le lac Érié, et Port Dalhousie, sur le lac Ontario, au moyen de quarante écluses de bois. Il amena la prospérité à la région en permettant aux produits du Haut-Canada de passer par New York. En 1841, le Gouvernement canadien entreprit la reconstruction du canal pour en accroître la valeur militaire et commerciale.

Historic Sites and Monuments Board of Canada.
Commission des lieux et monuments historiques du Canada.

7.10 Allanburg, 1989: When this community (where the first sod was turned for the Welland Canal in 1824) was surveyed in 1836, considerable growth was anticipated [see Chap. 6, Fig. 6.7]. By 1851, with about 300 inhabitants, it could boast of two grist mills, two saw mills, two woollen factories and a tannery. In 1876 the population was down to about 200, but it still advertised its "good milling facilities" on the Welland Canal. By the late 19th century (with a population of only about 100), it was described as having "good waterpower . . . not utilized." Today little remains, but there is a commemorative cairn. – R.R. Taylor; Smith, *Canada, Past, Present and Future*, 1851; Page's *Historical Atlas for the Counties of Lincoln and Welland*, 1876; *Ontario Gazetteer and Business Directory*, 1884-5

7.11 Port Robinson, ca. 1920: In this quiet, bucolic scene, Heslop's flour and grist mill, located where a cut from the Second Canal connected with the Welland River (Chippawa Creek), is a forlorn reminder of the once thriving international port. In its heyday in the latter part of the 19th century, it was home to at least two ship builders (one with a dry dock in the old wooden lock), several flour mills, a broom factory, and a carding mill. Successive rebuilding of the channel wiped out industrial sites, while flour milling was attracted to Port Colborne, and lumber mills ran out of local supplies. – AO: Pamphlet 1920, no. 7

7.12 Welland, 1958: Welland's coat of arms honours the canal shipping upon which, until recently, the community depended. Construction of the By-Pass in 1973 removed canal traffic from the heart of the city. Bridge 13, a characteristic lift bridge, which replaced the Alexandra Bridge in 1929, is still a popular landmark, although it ceased to operate in 1972, when the By-Pass was built [see also Chap. 12, Fig. 12.6]. – Welland Public Library

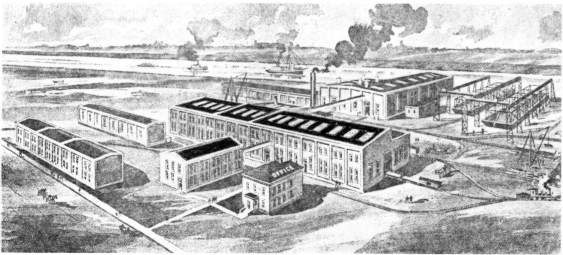

7.13 Welland, 1907: The 200 workmen of this foundry (established in 1826 by Matthew Beatty, an Irish emigrant) produced contractors' equipment, including dredges. Canal-stimulated industries such as Beatty's (demolished only in 1989) have continued to provide vital products and employment in Welland, serving not only the Niagara Peninsula, but other areas as well. – *Welland Telegraph*, 1907; *The Canadian Album: Men of Canada*, Vol. 1. Brantford, 1891

7.14 Wainfleet, 1989 (left): The Feeder Canal, once the town of Marshville's lifeline to the wider world, is now overgrown (foreground). The general store (left) and the hotel and postmaster's house (right), as well as a feed store, and a few other shops, are all that remain of the village's early prosperity. – R.R. Taylor

7.15 Port Maitland, ca. 1910 (above): By the late 19th century, when the Feeder was ceasing to be an active commercial artery, Port Maitland began to develop into a resort community. The Exchange Hotel flourished for many years under a variety of names. The village is still a popular summer location for boating and swimming. – Burtniak Collection

7.16 Port Maitland, 1989 (left): Weedgrown and abandoned, the Feeder Canal lock is no longer in use [see Chap. 6, Fig. 6.17], but some industry remains — Powell's Shipyard, for example [see Chap. 3, Fig. 3.10]. The concrete cribs used in construction of the Port Colborne piers were built here in 1926 and towed along Lake Erie the following spring. – R.M. Styran

7.18 Dunnville, ca. 1920: The elegant Victoria Hotel on Chestnut Street suggests the prosperity which Dunnville retained for many years, despite the closing of the Feeder and the failure of Grand River Canal ventures. – Burtniak Collection

7.17 Dunnville, 1890s: Typical advertisements placed in the directories testified to Dunnville's continuing prosperity. Today many of the earlier factories and businesses have gone, but the community is still a busy local centre. – Morrey's *Directory for the Counties of Brant . . . Lincoln, Welland . . .*, 1888

7.19 Port Colborne, ca. 1890: In this prosperous canal-town scape, stores and businesses line the channel, facing the source of trade. West Street stores overlook the harbour, the Second Canal lock (where the gates are shut), and the Third Canal lock (with its gates open). A typical swing bridge leads to East Street (out of sight, left). This was the home town of DeWitt Carter (1849-1933), a master mariner, an entrepreneur (such as Noah Phelps in Merritton or Thomas Rodman Merritt in St. Catharines), and mayor of the city. – AO: 13967-13; PCHMM

7.20 Port Colborne, ca. 1920: By the end of Carter's life, in the Fourth Canal era, Port Colborne had developed a large harbour and a number of associated industries. Coal is being loaded at the left, and several dredges are at work enlarging the harbour for the Fourth Canal. In the background are the Maple Leaf Mill (left), rebuilt after an explosion in 1919, and the government grain elevator (right). Flour milling is still a key industry in Port Colborne and the huge mill and elevator are visible symbols of the role the Welland Canals have played in the city's history. – Burtniak Collection

Chapter 8: Creating Employment

8.1. Thorold, ca. 1882: The canals have always provided a variety of jobs, skilled and unskilled, yet we lack accurate visual documentation of the earliest of occupations connected with the First Canal. No drawings of construction and few of the towboy, that most essential worker once the canal was completed, have survived. Contrary to the romantic image portrayed by this artist, the towboys' work was dirty, low-paid and boring, even in good weather. These men – rarely "boys"– kept the ships moving on the First and Second canals. By the late 19th century, when the Third Canal was built, towboys were being superseded by steam tugs. – SCHM: N-3678

There was nothing but mud, up to the knees in some places, for the poor fellows on shore who wearily plodded along. . . . – A description of the Second Canal towpath, above Thorold; David Cowan. – *A Report on Canal Navigation in Canada*, 1870

8.2 Irish Emigration, ca. 1840: Desperate poverty drove thousands of people out of Ireland; many of them fell into a similar situation as labourers on the Welland and St. Lawrence canals. Not even the most romantic artist in Niagara found their plight worth recording, as they endured disease (malaria, cholera, typhoid), long working hours, inadequate housing, and ethnic and religious prejudice. Not having the means or opportunity to become farmers, they endured seasonal unemployment, and with their wives and children they moved from one construction camp to another, thus lacking even a stable home life. – NAC: C-3904

LABORERS
WANTED ON THE
Welland Canal.

A**S the Company are determined to finish this Canal the present season, good encouragement will be given to all classes of Labourers: $12 per month will be paid to common *shovellers*, with a privilege which will render it not difficult to obtain $15. Good, active, smart men, as teamsters and men to hold the plough, can have from $15 to $16, with a chance of earning and receiving more: smart, active men, who are capable of keeping the time and overseeing 25 men, can have $20 per month, with a chance of extending their wages to $25. Any person that will bring on to the work two good yokes of oxen and a good stout cart, shall receive $28 per month, and himself and team found; and can have employ until the Deep Cut is completed. Any person employing and bringing on fifteen good shovellers, shall be entitled to the wages of an overseer, and hold that station, and may draw for his men the wages above stipulated; or seven cents per yard for shovelling into carts or wagons, (after the earth is ploughed up,) and his own pay as an overseer. *All those wishing employment in any of the above situations, will find it their interest to apply immediately.*

All persons employed on this work may be assured, that good Rules and Regulations will be adopted; and, although the country is generally healthy, still sickness is more or less prevalent in all places: Therefore, the Subscriber will erect a suitable House for the accommodation of the sick, where all necessary Medical aid will be administered to the labourers, *gratis*, together with all other attention that the nature of the case may require. All classes of MECHANICKS will do well to visit the WELLAND CANAL, and judge for themselves of the encouragement offered.

☞ Application may be made to the Subscriber, who has contracted to finish the Deep Cut, and is concerned in building all the Locks, and who will generally be found at the Deep Cut.

N. B. Cash will be paid for 100 yokes of good, young, working Oxen, and for all kinds of Grain.

OLIVER PHELPS.

Centreville, May 28, 1827. 71

P. S. It is desirable that several convenient BOARDING HOUSES should be opened in the vicinity of the cut. $1 50 per week will be paid for good common board and lodging, during the progress of the work.

8.3 The Realities of Immigrant Life, 1827:

So soon as the cholera made its appearance, the frightened workmen fled from the scene of death. One doctor fell a sacrifice to his humane exertions; a second, hired at double pay, was seized with sickness, and obliged to return to St. Catharines. – Annual Report of the Directors of the Welland Canal Company for 1832. *St. Catharines, 1833*

Heard that the cholera had commenced its ravages that day at Gravelly Bay – three deaths. Went on to the Bay that evening and found Coondor, a Contractor . . . with a man by the name of Henry, working on the lock, and one Ross, a labourer, at the same place, was dead and taken only that morning. Three or four were considered dangerous. Only one has died since, the others recovered. W.H. Merritt to his wife, 21 July 1832, Biography

No unions or social security existed in the First and Second Canal eras.

JOURNAL.
St. Catharines, December 22, 1842.

Canal Rioters.—We regret to learn, that there has lately been some further disturbances among the laborers on the Feeder of the Welland canal, in the vicinity of Broad Creek. The troubles, however, do not appear to have been extensive—the unemployed men, who attempted to stop the progress of the work, being so few in number, that, on the appearance of Baron de Rottenburgh, the special Magistrate, backed by the troops, stationed at Dunnville, the leaders were taken into custody, and the rabble speedily dispersed.

8.4 St. Catharines, 1876: The General and Marine hospital, founded by Theophilus Mack in 1870, on Queenston Street overlooking the Second Welland Canal, served injured or ailing sailors as well as local people. The Welland Canal Mission, founded in 1868, to care for the spiritual well-being of mariners, still functions, with its headquarters in St. Catharines. – J. Lawrence Runnalls, *A Century with the St. Catharines General Hospital.* St. Catharines, 1974; Welland Canal Mission to Sailors

8.5 Broad Creek, 1842: That the journalist could label the starving unemployed "rabble" indicated local attitudes to the labouring poor. – *St. Catharines Journal,* 22 Dec. 1842

PUBLIC WORKS OF CANADA.

No. of Pay-list *2261*

Pay-list in duplicate of persons employed on the (1) *Welland Canal*
during the Month of (2) *March 1873* (3) to *Repairs of Banks, Locks & Bridges &c &c.*
chargeable to (4) *Repairs*

No. of Letter authorizing Salary and Wages returned.	CAPACITY.	When Employed.	NAMES.	TIME.	RATE.	AMOUNT.	I acknowledge to have received the sum set opposite to my name and signature in payment of service rendered as entered in the Pay-List.
	Carpenters	March 1873	James Dell	26 days	$ 2.50	65 00	James Dell
	"	" "	James Ferres	25 "	1.75	43 75	James Ferres
	"	" "	John Henning	19½ "	" "	34 12	J Henning
	"	" "	William Jones	1 "	2.00	2 00	William M Jones
	"	" "	James Collier	26 "	1.50	39 00	James Collier
	"	" "	James Delaney	25 "	"	37 50	James Delaney
	Hauling Material	" "	Chas Hill + team	24½ "	3.00	73 50	Chas Hill
	Masons	" "	John Pocock	9½ "	2.50	23 75	John Pocock
	"	" "	William Pocock	11½ "	" "	28 75	Wm Pocock
	"	" "	Thomas Pocock	9½ "	" "	23 75	Thos Pocock
	Laborers	" "	John Delaney	26 "	1.25	32 50	John Delaney
	"	" "	Alx Winslow	6 "	" "	7 50	Alx Winslow
	"	" "	Pat McCoy	6 "	" "	7 50	Pat McCoy
	"	" "	James Bradley	25 "	1.50	37 50	James Bradley
	"	" "	Birney Clark	24/4 "	" "	36 37	Birney Clark
				Carried forward		$ 492 49	

8.6 Welland Canal Paylist, 1873: Repairs to the locks, banks and bridges were continual and required a regular workforce of carpenters, masons, teamsters and unskilled labourers. The continuing Irish presence is evident in the names: the Bradleys, for example, were a dynasty of canal workers. Repeated names, such as the Delaneys and Pococks, suggest either a certain solidarity among these men or the scarcity of other jobs for relatively unskilled labour. The best-paid man is one with a team. Many labourers acquired some skills, and as they settled in canal communities, the developed a pride in their accomplishment and their home towns as well. – SCHM: Welland Canal Files

8.8 Fourth Canal, ca. 1925: White-collar workers were also necessary to the construction of all the canals, but we lack any pictorial record of such engineers and office staff for the earlier Welland Canals. Among the many clerical jobs would be the task of keeping track of reports from the field [see Fig. 8.9]. – SCHM: N-6681

8.7 Welland, ca. 1880 (opposite): Workers, surrounded by mud, are dwarfed by the giant stone structure of the aqueduct. Self-respect must have been hard to maintain, working in these dangerous, filthy conditions, even though a certain degree of skill was needed for many of these jobs. Note the still-essential shovel. Samuel Woodruff (1819-1904, insert, left) was a sober and ambitious civil engineer who was involved in construction of the Second Canal and later became canal superintendent. William Pay (1819-1904, insert, right) emigrated, alone, from England to Canada when only 16. A practical, proud, hard-working carpenter, he became car and track superintendent of the Welland Railway, which serviced Third Canal construction and workers. – WHM 988.022.003.002; Woodruff, *The Canadian Album: Men of Canada, Vol. 5. Toronto, 1896;* Pay, SCHM: N-3611

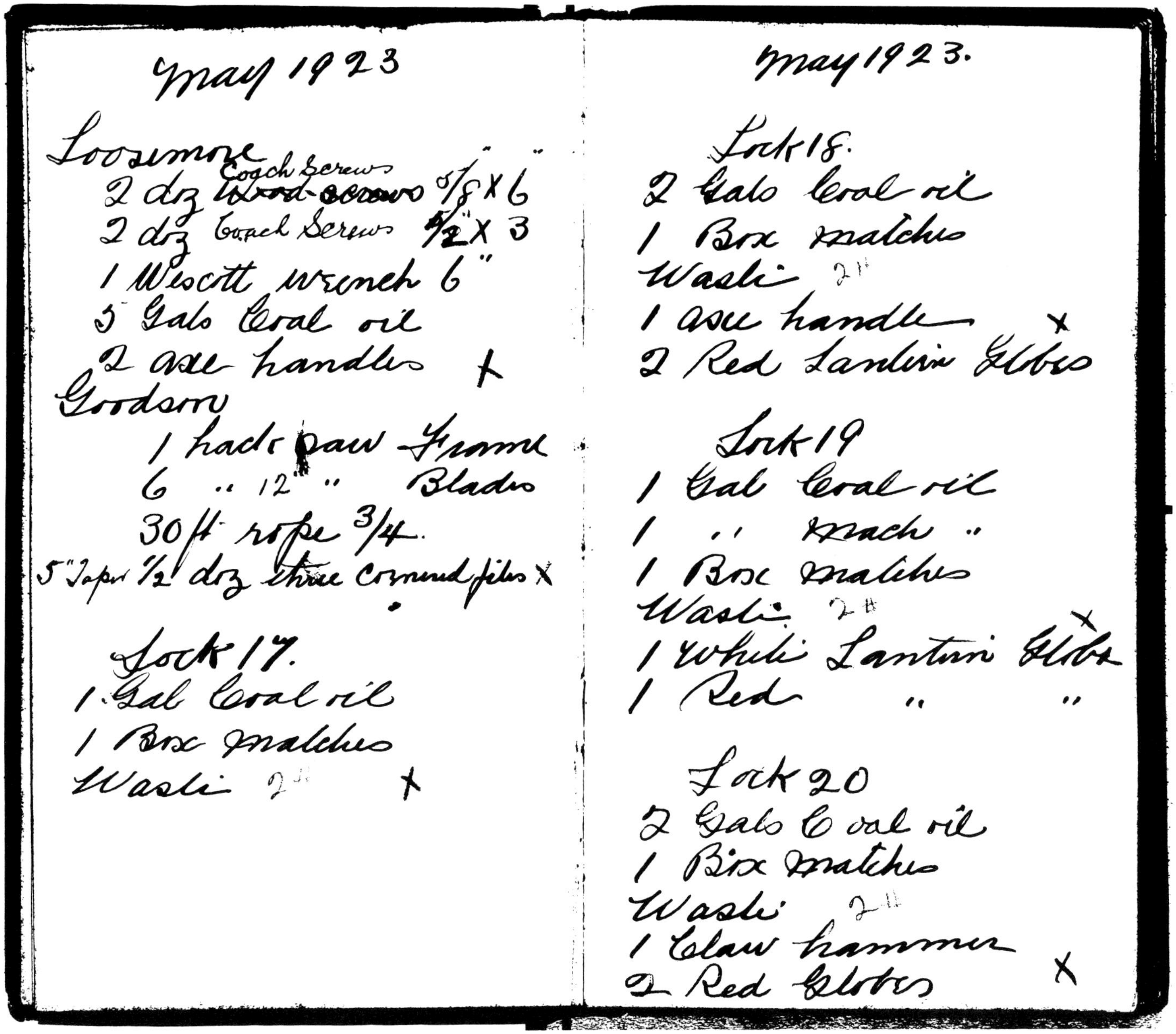

May 1923

Loosemore
Coach Screws
2 doz Wood screws 5/8 × 6"
2 doz Coach Screws 1/2" × 3
1 Wescott wrench 6"
5 Gals Coal oil
2 axe handles ✗

Goodson
1 hack saw Frame
6 " 12" " Blades
30 ft rope 3/4"
5" Taper 1/2 doz three cornered files ✗

Lock 17.
1 Gal Coal oil
1 Box matches
Wash 2# ✗

May 1923.

Lock 18.
2 Gals Coal oil
1 Box matches
Wash 2#
1 axe handle ✗
2 Red Lantern Globes

Lock 19
1 Gal Coal oil
1 " matches "
1 Box matches
Wash 2#
1 White Lantern Globe ✗
1 Red " "

Lock 20
2 Gals Coal oil
1 Box matches
Wash 2#
1 Claw hammer
2 Red Globes ✗

8.9 Third Canal Supply Book, 1923: While the Fourth Canal was under construction, the Third Canal had to be kept in operation. These pages from Frank Hoover's supply book are typical of on-site reports required by the government. Something of the texture and smell of day-to-day labour is still evoked by this list of supplies needed by the maintenance worker to carry out his duties. The coal-oil, presumably for lamps, suggests a night shift or perhaps lack of illumination in some areas. – SCHM: Welland Canal Lock Company, Minute Book

8.10 Fourth Canal Surveyors, ca. 1920: The crews who surveyed the canal routes had to work in all kinds of weather and terrain before any construction could proceed. This group was obviously proud of their contribution! Their equipment was more sophisticated than that used a century before, but their fundamental role in construction and maintenance of the canal had not changed. – SCHM: N-7239

8.11 Port Dalhousie, ca. 1913: Yet another type of essential worker on the canals was the lock tender. Toby Johnson and Scotty Nichol pose proudly in front of their shanty at Lock One of the Third Canal. The building can still be seen, the last one to have survived [see also Chap. 12, Fig. 12.8]. – SCHM: N-1792

Welland Canal,
   Office of
Superintending Engineer.

          St. Catharines, Ont., September 4th, 1900.

Mr. M. Henry,

     Port Dalhousie, Ont.

Dear Sir;

     M. McCarthy, Lock master at Lock 25, informs me that since you fixed the time for himself and Nelson Higgins to exchange watches at Lock 25 that he has never been more than three minutes late, and that he is of opinion that Nelson Higgins' watch is not reliable.

     Please enquire into this question of time at Lock 25. The Standard time is, of course, what must govern.

     McCarthy also informs me that Higgins is not disposed to recognize him as Lock master.

     I should like to have a full report from you on this subject as soon as possible.

             Yours truly,

             W.G. Thompson

             Superintending Engineer.

– SCHM: Norris Papers

8.12 Third Canal, ca. 1900: This picture is probably of members of Merritton's Boyle family. The men seem to have posed for a special occasion, judging from the proud and self-conscious posture of the four on the box. The well-dressed boys at the right were likely spectators. They may all have worked on what seems to be a floating derrick or dredge. Photos like this can be found in many family albums and attics throughout Niagara. – St. Catharines Centennial Library

8.13 Third Canal, ca. 1920: The BENMAPLE, a ship of the Matthews Line, reminds us that the ships which transit the Welland Canals have provided numerous jobs. Seamen's costumes have changed over time, but the skills required to manoeuvre ships through locks have remained complex. Captains and pilots still must be familiar with every lock and bend of the waterway, as they have from earliest days on the canal. – SCHM: N-1824

8.14 Near Port Dalhousie, ca. 1908: The tug ESCORT sank in Lock 1 in November 1907, after a collision with a barge, drowning her captain and two others. Lock 3 was used as a temporary drydock for the salvaged ESCORT – an example of winter employment on the canal. – SCHM: N-4866

A. V. STAATS,

ICE, WOOD AND COAL

83 GENEVA STREET,

ST. CATHARINES, ONT.

TELEPHONE 256.

8.15 Port Dalhousie, ca. 1913 (left): When the canal was shut down for the winter it provided ice for local communities, in the days before refrigeration became common. This burly ice-cutter takes a rest while his partner wields the tongs, providing an essential ingredient for the business whose sign appears in the background. The men probably heaved coal and wood in the winter months, as well as ice in summer, as it was common practice for dealers to combine these two seasonal trades.
– SCHM: N-1767; Vernon's *St. Catharines Directory for 1906-07*

8.16 Port Dalhousie, 1900 (opposite): Group photography was still novel enough to draw out most of the employees of the Maple Leaf Rubber Works. The occasion may have been the reopening of the factory after a fire of that year. Maple Leaf Rubber had opened in 1890 in the former Lawrie Flour Mill: an example of how canal-side mills often changed ownership and function, and were renovated and expanded. Such changes often occurred after fires, but the mills continued to provide work – here, for a small army of men and boys.
– SCHM: N-7470; Fuller's *St. Catharines Directory for 1863-64*

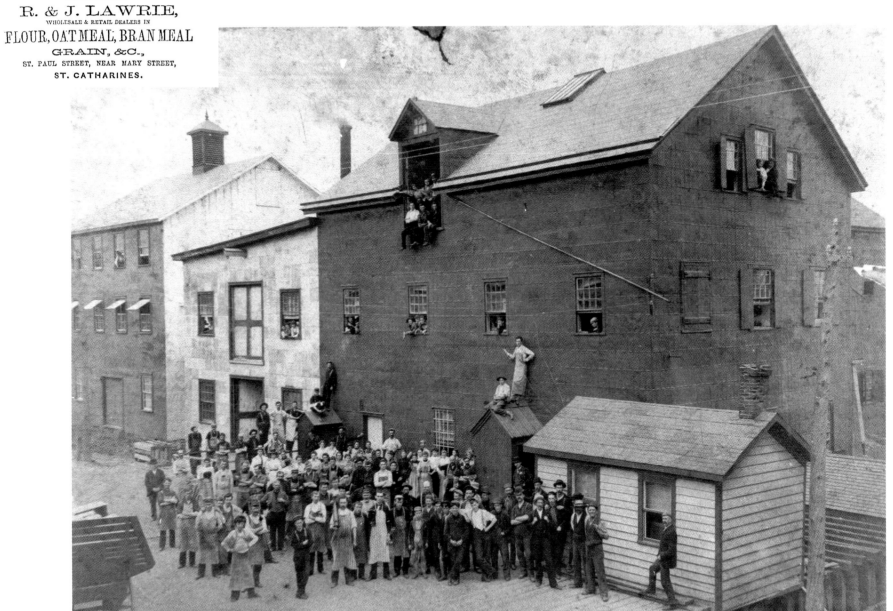

Chapter 9: People And Pleasures

9.1 Port Colborne, ca. 1910: Since earliest times, ships and water have provoked a sense of awe, making ship- and canal-watching one of the pleasantest of free entertainments. For over 160 years the Welland Canals have provided countless hours of recreation for many people. Much of the fun is casual, as here, on the west bank of the harbour, where local boys look out to Lake Erie, watching for ships.
– PCHMM: 980.60.ll

9.3 St. Catharines, ca. 1982: Ships themselves have always been a major attraction. Even a work-worn laker or a grimy salty attracts shipwatchers. But Lock 3's viewing stand offers two sights — a close view of the lock in operation, as well as of the passing ships. Every summer, thousands marvel as gravity and the weight of water lift or lower tons of steel and cargo. Onlookers exchange greetings with seamen from Hong Kong, the Pireus, Melbourne. . . . – R.R. Taylor

9.2 Port Colborne, ca. 1900: The boys above might have just spied this yacht entering the harbour. The Third Canal breakwater was a challenge for those who enjoyed a brisk "constitutional," although perhaps only in favorable weather! The fence on the left was designed to offer the lighthouse keeper a handhold as he approached or left the light during storms. The grain elevator of the Grand Trunk Railway remained a prominent landmark until 1911. – PCHMM: P-977.719

9.4 Welland, 1964: The passage of the CHRISTIAN RADICH through the canal drew crowds of viewers, both locals and tourists, because this type of ship, once a commonplace sight, is now a rarity. It is seen here passing under Bridge 13.

– WHM

9.5 St. Catharines, 1988: Ship-watchers can get a close-up of the bridge of a laker, the FORT HENRY, retired by Canada Steamship Lines in 1979 after 35 years on the lakes. The contradiction here is more apparent than real: the bridge was bought by Fortune Navigation of St. Catharines, who operated it as a tourist attraction during the summer of 1988. The company hopes to recreate the top deck and use the bottom deck as a shop for quality marine art and artifacts. In the background the MONTREALAIS enters Lock 3. – R.R. Taylor

9.7 Port Robinson, c. 1910: Some pleasure-seekers plunged right into the water! The waterway continued to attract "canal rats," who engaged in an increasingly dangerous pastime. The old port was still an operating harbour as the presence of the tug and the DREDGE CHIEF shows. – Petrie Collection

9.6 Wainfleet, ca. 1910: Even abandoned parts of the waterway draw pleasure-seekers and sportsmen. Here, the raceway of the Feeder Canal has lured fishermen to its banks, but no doubt others strolled down to the water for a few moments of "peace and quiet," to listen to the frogs, or perhaps to catch sight of a rare bird. – Burtniak Collection

9.8 Port Dalhousie, ca. 1910: Another focus for recreation was Lakeside Park, developed by the Canadian National Railway in 1902. It flourished every summer until the 1950s. The carousel, later enclosed to protect it from the elements, was a big attraction. Today, carefully restored and housed in a new building, it is the only surviving ride in the park – still offering rides for only a nickel! The electric Niagara, St. Catharines & Toronto Railway (N.S. & T.) is in the foreground. – Burtniak Collection

9.9 Welland, 1976: Less risky than swimming in the canal itself [see Fig. 9.7] was a dip in a swimming pool created in the disused Second Canal aqueduct. The "Cross Street Pool" was opened in 1946 and closed only in the late 1980s. – R.R. Taylor

9.10 Port Dalhousie, ca. 1927: If one tired of the beach, the pedal-boats, the water slide, the carousel, or the games of chance, there was always "the Falls" – accessible via the N.S. & T. – SCHM: 988.33

9.11 Port Dalhousie, June 21, 1958: Until 1950, regular ferries such as the GARDEN CITY ran between Toronto and Port Dalhousie, and to other ports on the south shore of Lake Ontario. The last large passenger ship to dock at "Port" was the LADY HAMILTON, linking Hamilton, Niagara and Queenston, and offering both daytime and moonlight cruises.
– SCHM: N-6247

9.12 Port Colborne, 1931: Other passenger ships travelled through the waterway itself. "Hello kids!" says the message on this postcard, "Well, here I am on the ship NORONIC and enjoying it fine." Shown here in front of the Maple Leaf grain elevator on its maiden voyage, the NORONIC, with six decks, was fitted out like a luxury liner. It carried 562 passengers and made regular transits through the Fourth Canal until it was destroyed by a disastrous fire at Toronto in 1949. – Burtniak Collection

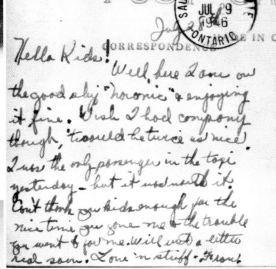

9.13 Port Weller, 1988: An old tradition was revived in 1988, when a new and smaller GARDEN CITY [see Fig. 9.11] began to make regular trips between Port Dalhousie and Port Weller, and occasionally through Lock l of the Ship Canal. Fortune Navigation hopes to run some day trips between Port Dalhousie and Lock 3, and at least one complete weekend transit of the whole canal. – R.M. Styran

9.15 St. Catharines, ca. 1890: Schoolboys have always found their own uses for the waterways. On the canalized Twelve Mile Creek, older Ridley boys played hockey in the winter; in the summer, they swam in or canoed on the stream. – Peter Gzowski, *A Sense of Tradition.* St. Catharines, Ridley College, 1988

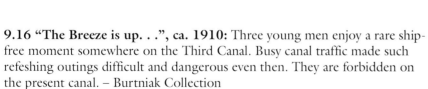

9.14 Welland Ship Canal, 1931: Educators have exploited the waterway for their own, less frivolous purposes. Here a group of fourth formers from Ridley College's Lower School enjoy a holiday from classes traveling through the canal. In the process, they learned about the great new waterway, local geography and international trade. School classes still visit the canal, but few are lucky enough to get aboard a ship. – SCHM: N-8380

9.16 "The Breeze is up. . .", ca. 1910: Three young men enjoy a rare ship-free moment somewhere on the Third Canal. Busy canal traffic made such refeshing outings difficult and dangerous even then. They are forbidden on the present canal. – Burtniak Collection

Water Sports at Welland, Ont.

9.17 Welland, ca. 1920: Some boaters could be more aggressive. Near the still new Alexandra Bridge, a modern motor launch disturbs the canoers. Abandoned by the Seaway in 1974, this stretch of the waterway is now given over entirely to water sports. – Burtniak Collection

9.19 Dunnville, 1989: The waste weir near the Grand River dam on the Byng side of the water is still part of a peaceful, attractive landscape, as it appeared to a reminiscent native in 1893. – R.R. Taylor

9.18 Port Dalhousie, ca. 1905: A more formal and skillful type of water sport has been practised since 1903, in the Royal Canadian Henley Regatta. This is the finish of a strenuous race before a large crowd of fans. – SCHM: N-1484

Like some old veteran, whose battles are over and whose bustling days are past, it has now that quiet charm which repose and decay alone can give . . . A pleasant, airy, picturesque spot it is. . . . the swift current sweeps into Sulphur Creek, and when the valves of the weir are open and the water is rushing through them, it becomes so rapid and strong that it would be rather difficult to stem it. When, like the "Sweet Afton," it flows gently along, the lake fish delight to come up and play here. On each side of the channel you may see one of the quaint-looking dip-nets used in this region, with its long balance pole and its upright rest. It hangs over the water ready for a dip whenever indications are favorable. . . . – [Quoted from "At the Mouth of the Grand," Canadian Magazine, July 1893.]

9.20 Welland, ca. 1900: Before air-conditioning, sweltering Wellanders were drawn to the breezes along the canal banks. Festivities were planned for nearby sites, as here on Canal (later King) Street. A band concert on the podium (left background) entertains a well dressed crowd on the Third Canal bank, on a hot summer afternoon, possibly Dominion Day. – WHM

9.22 Port Weller, ca. 1930 (opposite): The builders of the Welland Ship Canal took the public's fascination with the waterway into account and planned to have parks and gardens along its length. This rock garden and greenhouse were part of a large nursery, itself to be an attraction. The Great Depression and the Second World War terminated the project. All that remains today are the rows of trees planted along the canal banks to act as windbreaks. – PCHMM: P-979.16.475

9.21 Port Dalhousie, 1941: The Picnic Committee of the Dofasco employees is a reminder of the many picnic groups drawn to Lakeside Park and "Port's" harbour sights. By mid-twentieth century, fashionable summer clothes were more comfortable and sensible than they had been in 1900! – SCHM: N-7325

9.24 Third Canal, ca. 1905: The boy in overalls, the son of a ship's captain, shows his landlubber cousin some of a sailor's skills on board a tug. Few canal-watchers are fortunate enough to actually board a passing ship today, since only larger motorized pleasure craft are allowed to share the waterway with the freighters and other canal work-horses. Those canal buffs lucky enough to travel the length of the Welland on a tug or laker can only be grateful for the opportunity to share with the crews the sights and sounds unobtain able from shore and admire the judgement and skills required to take a laker through the locks. – AO: Murphy Collection - 4

9.23 Welland Ship Canal, ca. 1928 (left): Construction sites on successive canals have had a special appeal, drawing the attention of many sight-seers. Here a family — probably that of an engineer — enjoys a privileged view of the inside of a lock from atop a safety gate. Increasing concern for public safety, and for the costs of liability in case of accident, have meant that public access to the vicinity of the locks has had to be curtailed. – PCHMM: 979.9.4

Chapter 10: Difficulties And Disasters

10.1 Port Robinson, 1974: The wooden locks of the First Canal were often out of repair, hindering navigation. Since the rebuilding in stone, in the 1840s, the canals have generally run smoothly. Construction and operation, however, have occasionally met with difficulties; less frequently, with major disasters. Simply living near the canal can create nuisances or even perils for residents. For canal operators, the worst calamity possible is the rare blockage of the waterway and the resultant halt to traffic. For example, Bridge 12 crumpled after it was rammed by the ore carrier STEELTON. Fifty vessels waited for passage for two weeks as cranes removed the debris. The little canal community was permanently divided, for the bridge has not been rebuilt. – Alfred F. Sagon-King

10.2 Port Colborne, September 1989: To avoid a repetition of the STEELTON disaster, the tug ARGUE MARTIN carefully guided the H.M. GRIFFITH towards the dock after a conveyor belt caught fire in the coal-carrying freighter. At the dock, firefighters eventually controlled the blaze. Traffic in the canal was held up for only four hours. – *The Standard*, St. Catharines, Ontario, 28 Sept. 1989, Mike Conley

10.3 Thorold, 1925: There have been construction accidents on all the canals. One of the most spectacular occurred on the Fourth Canal, when a Blaw-Knox travelling form for placing concrete collapsed killing eight workers. The accident happened while the form was being moved from the site of Lock 4 to that of Lock 5. Note the tiny figures at the base of the lock wall, examining the damage. – Alfred F. Sagon-King

In Loving Memory of little

Willie Wallace,

who was accidentally drowned on June 20, 1912,

at Lock 21, New Welland Canal,

Thorold, Ontario, Canada,

aged 5 years.

We do not
We did no
We only kn
And never

Day by day
Words wo
Our loss is
In Heaven

Short was th
But peacef
And grandma miss
Because she loved you best.

10.4 Thorold, 1912: Despite the pleasure that the canals have given to so many over the years, the force of their dammed-up waters has always been dangerous. Swimming, fishing or even simple ship-watching can be fatal. Three innocent bystanders were drowned when the government ship LA CANADIENNE smashed through Lock 22's gates after experiencing a mechanical failure. – SCHM: 982.233.2

Niagara

Niagara, lately the district town of the Niagara District, and now the county town of the united counties of Lincoln and Welland, was formerly called Newark. It is one of the oldest settlements in Upper Canada, and was for a short time the capital of the country. It was once a place of considerable business, but since the formation of the Welland Canal, St. Catharines, being more centrally located, has absorbed its trade and thrown it completely in the shade.

– Smith's *Canadian Gazetteer* (1849)

Queenston

A village in the township of Niagara, seven miles from the town of Niagara, seven miles from the Falls, and forty-seven miles from Hamilton. It is pleasantly situated on the Niagara River, below the Heights, and opposite the American village "Lewiston." Before the opening of the Welland Canal, Queenston was a place of considerable business, being one of the principal depots for merchandise intended for the west, and also for storing grain: as much as forty or fifty thousand bushels of wheat have been shipped here in a season, which now finds its way by the Welland Canal. – Smith's *Canada, Past, Present and Future* (1851)

10.5 St. Catharines, 1826: While the waterway's success proved a boon to the economies of canal corridor towns, it spelled stagnation or decline for some other Niagara communities. Towns such as St. John's and Niagara (now Niagara-on-the-Lake) lost businessmen who removed their enterprises to more profitable locations in the canal ports. The adverse effects on both Niagara and Queenston were noted in mid-century publications. – *Farmers' Journal and Welland Canal Intelligencer*, 28 June 1826

C.Whitten J.Houghton D.Dewey G Moore R Hodgson L.Smith S.Anderson
C.Smith B. Roham F.Haney W.Scott C.Helfrick F. Hodson C.Harper
1920

10.6 St. Catharines, 1928: Even when canal construction was over, problems could develop. For example, Haney's Garage (founded by Jack Haney, who had driven a REO automobile coast to coast in 1912) had assembled and maintained large construction machines used on the Fourth Canal. When building was completed in 1932, business suffered. The Depression did the rest and the garage closed in 1935. – SCHM: N-5397

10.7 St. Catharines, 1869: Throughout its history, the canal system has had to be frequently repaired, in an attempt to maintain it in good working order. A survey for an "improved" canal, prepared in 1837, noted that, regarding the First Canal locks at the Niagara Escarpment, numbers 20 to 24, as well as 28 and 30, were "out of repair." The continuing problem is illustrated by an incident reported in the *True Patriot Paper* (10 September 1869):

On Monday morning last, a break took place in the waste weir, above Lock 2, and since that, the navigation on the Welland Canal has been at a standstill, and much disappointment is being expressed by skippers, shippers, and owners at the way things are being managed by Mr. . . . [sic]. But whether this break in the waste weir might have been anticipated by the Superintendent observation is a matter we know nothing about, but this we do know that several of the Locks are sadly the worse for wear, and that a very slight accidental pressure by any of the lightest crafts that sail the canal, might subject the owners very probably to $800 damage as has been done heretofore, for want of proper repairs, which is nothing uncommon . . . .

10.8 Thorold, 1985: (Opposite) Even the mammoth concrete walls of the Fourth Canal, some over sixty years old, need rebuilding. When a 45.7 m (150 ft) section of the concrete wall on the west side of Lock 7 collapsed, a ship was trapped, and traffic was held up for twenty-three days near the end of the season, while repairs were carried out. This accident led to a thorough examination of all the locks and to a seven-year repair programme estimated to cost $175 million. The program will completely rehabilitate the "middle-aged" waterway.
– SCHM: N-6175, *St. Catharines Journal*, 1 June 1837

10.9 St. Catharines, ca. 1900: Lock gates, as well as the lock chambers, require attention. Here the latest underwater technology was used to check the condition of Third Canal gates at Lock 7, near Niagara and Vine streets. While a diver prepares to descend into the murky, dangerous water, assis tants are ready to hand-pump his air supply to him through a hose. An underwater examination of a temporary dam at Lock 1 of the Fourth Canal resulted in the death of two Seaway divers, on 30 January 1989: the first fatalities among the diving squad in the 30 years of opera- tion by the Seaway. – SCHM: N-1067

Lifting a Locomotive From Welland Canal Port C

– *St. Catharines Journal,* June 1837

10.10 Port Colborne, 1911: Human negligence on or near the waterway has caused a number of accidents. Several times in the canal's history, locomotives have failed to stop, smashed through barriers, or derailed, ending up in the channel. Even today, automobiles often have to be lifted from the depths of the waterway, having found their way there accidentally or deliberately. – PCHMM

10.11 Thorold, 1935: A potential threat to canal traffic comes from water pollution, such as the sawdust revealed here when the Second Canal was drained. In an attempt to prevent such difficulties, the Dominion Department of Public Works and mill-owners entered into complicated lease agreements regarding the Second Canal, one clause in an 1870 lease read: "Twelfthly, that no bark, chips, slabs, edging, sawdust, or other rubbish or refuse of any kind shall be allowed to fall or to be thrown from the premises hereby leased into the said canal or approaches thereto. . . ." The explicit penalty was official seizure of the mill. – St. Lawrence Seaway Authority: Western Region; Board of Public Works, *Register B, 1841-1880*

10.12 St. Catharines, 1837: Human malice has occasionally been as detrimental to smooth operation of the canal as mere carelessness. William Lyon Mackenzie's rebellion in 1837 made shipowners and shippers wary of entrusting vessels and cargoes to a canal in an area where several skirmishes had been fought.
– *St. Catharines Journal,* 26 April, 1838

10.13 Thorold, 1900 (above): Since its earliest days the canal has had strategic military importance because the Canadian and American economies depend on it. When a hole (bottom left) was blasted in one of the gates of Lock 24 of the Third Canal, on the brow of the escarpment, Fenians were suspected of having planted the dynamite. The Fenians, who supported Irish independence from Britain, were attempting to cripple both the waterway and the economic life of one of Britain's dominions. Fortunately, while the lock gates were blown open, the weight of the water closed them again, so that minimal damage occurred. – SCHM: N-1398

10.14 Port Robinson, ca. 1916 (left): Fears of further United States-based anti-British interference with the canal led to military detachments being posted to guard it during World War I. These soldiers are at the Second Canal lock leading to the Welland River [see also Chap. 11, Fig. 11.3]. Some First World War veterans became special officers in the RCMP to guard the canal in the Second World War. – SCHM: N-6841

10.15 Fourth Canal, ca. 1925: The very water which is a necessity for the operation of the canal can cause difficulties by appearing at the wrong time, as with heavy rain or seepage. This steam shovel, owned by the Dominion Excavating Company, is bogged deep in mud. – SCHM: N-7457

10.16 Port Colborne, 1939: Storm-blown waters have caused difficulties at both ends of the canal, and in its frozen form, water can inhibit canal traffic. Ice may have been involved in the sinking of the NORTHTON in 1939, and it certainly caused delay of navigation in earlier years [see Fig. 10.12]. – Welland Public Library

Drought stunts canal tonnage

By VINCE RICE
and PAUL HARVEY
Standard Staff

The devastating summer drought on western Canadian wheat farms has taken the steam out of the St. Lawrence Seaway's forecast for an improved year.

Seaway economists had estimated 45 million tonnes of cargo, mostly wheat, would be carried through the system in the 1988 season, a five-per-cent increase over last year, spokesman Harley Smith said today.

Now less than 41 million tonnes are projected, about the same as last season.

10.17 St. Catharines, 1988: On the other hand, the *absence* of water, even in other parts of Canada, can also create problems for the canal. – *The Standard*, St. Catharines, Ontario, 11 Aug. 1988

CANNING SEASON

10.18 St. Catharines, 1985: Almost since the day the First Canal opened in 1829, the increasing size of ships has created pressure for enlargement and has resulted in successive rebuildings of the waterway. Today the largest ocean-going ships are now too large for the locks: the "super-tankers" are about 610 m (2,000 ft) in length; the Welland Canal locks can take vessels of only 222.5 m (730 ft). Even the ships plying the Upper Great Lakes, at 305 m (1,000 ft), cannot come through the Welland into the lower Seaway, whose locks were built to similar dimensions. – *The London Free Press*, M.R. Tingley

Chapter 11: Changing the Landscape

11.1 Port Colborne, ca. 1870: Through most of the 19th century the tall masts and riggings of sailing ships in transit through the first three Welland Canals were prominent features of the Niagara Peninsula. They, and the belching stacks of the steam-powered vessels which superseded them, are rare sights today.
– PCHHM: P-979.754

11.2 Port Robinson, ca. 1972: Since the 1920s we have become accustomed to the lift bridges of the Fourth Canal dominating the Peninsula scenery. These bridges are an immediately obvious feature in a landscape which has been profoundly influenced by the successive versions of the Welland Canal. The bridge here was demolished in 1974, but the rest of the original eleven can still be seen. However, the five in the Dain City-Welland stretch of the canal have not been in regular operation since the opening of the Welland By-Pass in 1973. In this photograph the terrain is once again undergoing alteration, as the By-Pass (upper left) is under construction. – St. Lawrence Seaway Authority: Western Region

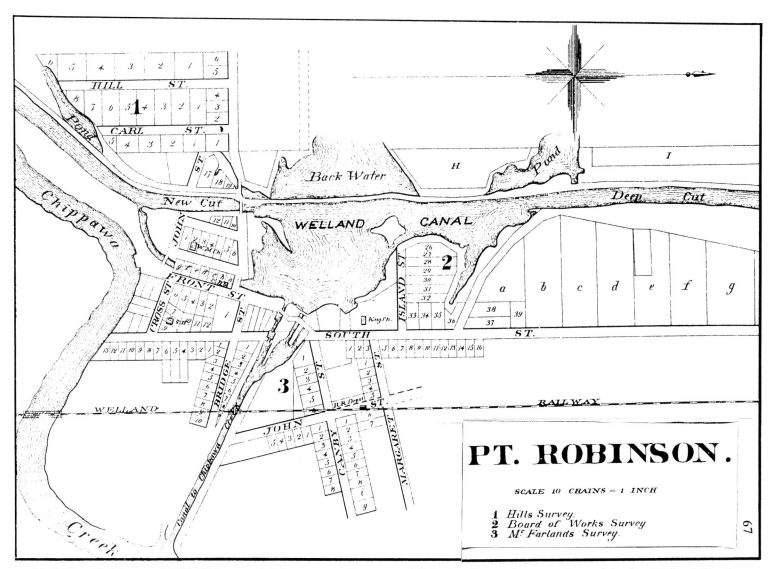

11.3 Port Robinson, 1875: This once thriving community was established at the southern terminus of the first Welland Canal, where the waterway joined the Welland River (Chippawa Creek). Even after the main canal line was extended to Lake Erie (1833), Port Robinson remained an important centre for passenger and freight traffic. This map illustrates how the first two canal cuts and associated ponds influenced the pattern of settlement. By the 1890s the port was in decline, and in the 1920s construction of the Fourth Canal further altered the landscape by obliterating the area shown here as the "New Cut." Today, only remains of the locks leading down to the Chippawa Creek can be seen . – Page's *Historical Atlas for the Counties of Lincoln and Welland*

131

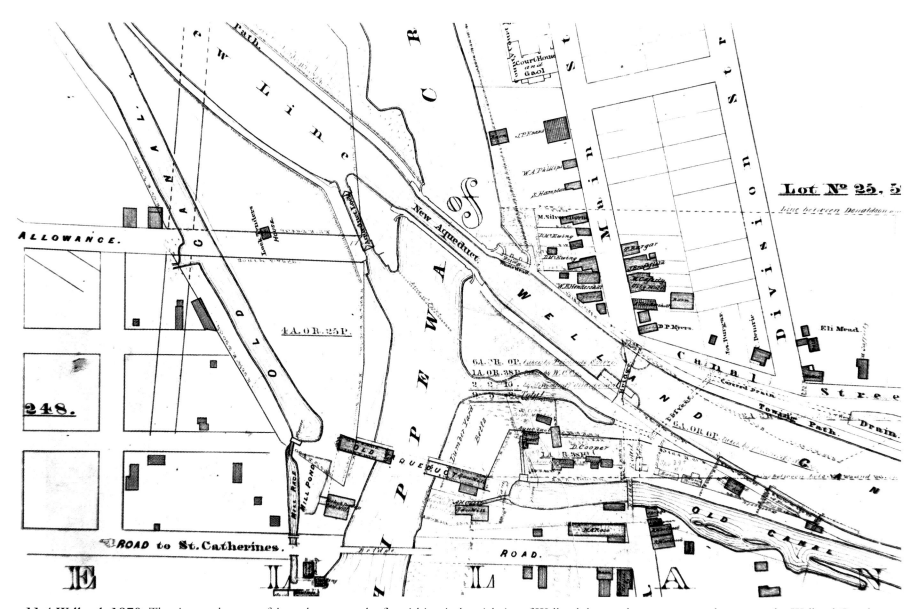

11.4 Welland, 1870: The tiny settlement of Aqueduct, now the flourishing industrial city of Welland, has under gone many changes as the Welland Canal has been rebuilt. Here the "new" aqueduct carries the Second Canal over the Welland River. The remains of the "old" aqueduct, the "old" (First) Canal, and the lock which allowed vessels to move between the river and canal, are on the left. The attraction of industry to the water power of both canals is obvious, as is the way in which streets and properties have had to adapt to the canal routes. – *Welland Canal Book 3*, undated, ca. 1870

11.5 Welland, ca. 1930: Comparing an aerial view with the 1870 map indicates further alterations to the landscape as the Third Canal aqueduct (III) continued to carry ships over the Welland River, while the syphon culvert for the Fourth Canal (IV) is under construction. The Second Canal aqueduct (II) is also visible here, and remains of it may still be seen [see also Chap. 9, Fig. 9.8]. Here, as elsewhere in the canal-side communities, the effects of the successive canals on the landscape can be seen long after the waterway has ceased to function as such. – P.J. Cowan, *The Welland Ship Canal Between Lake Ontario and Lake Erie, 1913-1932*. London, 1935

11.6 Welland, ca. 1910: The cut stone lock chamber on the right, with its typical wooden gates and nearby locktender's shanty, was a familiar aspect of the Welland landscape from the 1840s through the 1880s. – Burtniak Collection

11.7 Welland, ca. 1970: Although there are no locks in this section of the Canal, the Welland By-Pass (completed 1973) represents not merely the latest, but also the most dramatic, change to the area landscape. The winding course of the Fourth Canal was straightened and shortened: from 14.6 km (9.1 miles) to 13.4 km (8.3 miles). It involved exca vation of some 50 million m^3 (65 millions yds^3) of earth, clay, rock and silt, and required a second syphon, to divert the Welland River under the new channel. A major revamping of the road (some 80.5 km or 50 miles) and rail network in and around Welland was also necessary, as were several miles of new gas, telephone, sewer and hydro lines. (Two of the highest transmission towers Ontario Hydro had ever built were constructed on either side of the channel.) Several docks used by local industries were replaced by a single one of 305 m (1,000ft). – St. Lawrence Seaway Authority: Western Region

11.8 Thorold, 1920: The landscape in this area has undergone many changes, too. In the foreground is Lock 24 of the Third Canal, with Lock 25 beyond. The extensive canal reaches between locks, and the accompanying weir structures and ponds curving as the canal climbs the Niagara Escarpment, carve their way through the rural landscape. Already the next major alteration has begun, with construction of the Fourth Canal, to the left. This view appeared in at least two sets of postcards, with different captions [see also Chap. 5, Fig. 5.4].
– Burtniak Collection

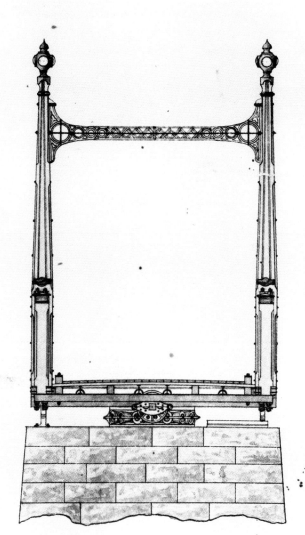

END ELEVATION
Double Track.

Canal and Locks
Dunnville, Ont.

11.9 Dunnville, ca. 1910: The Feeder Canal, running from the Grand River at Dunnville to join the main line at Welland, also affected the landscape across the Peninsula. This rather romantic view shows the arrow-straight Feeder disappearing into the distance, while the busy industrial town flourishes near the lock and the water power made available by the fall of water. – Burtniak Collection

11.10 Third Canal Bridge, 1880: The road and rail bridges carrying traffic over the canal became a notable feature of Peninsula scenery in the last quarter of the 19th century. The decorative detail on these utilitarian structures was not generally appreciated — and all have been destroyed long since. – NAC: NMC-21823

11.11 and 12 St. Catharines, 1915: While lift bridges are the most conspicuous of the Fourth Canal landscape features, other bridge designs were utilized. The view on the left could be a Roman aqueduct in southern France or Spain. In fact, it is the first Glenridge Bridge, built in 1915, when the Second Canal (which it spanned) was still in use as a source of hydraulic power [see also Fig. 11.13]. The steel span of the Burgoyne Bridge on Ontario Street (below), which connects the heart of St. Catharines with the "Western Hill" area, seems more modern, but was built barely a year later. This latter site has seen many changes: Shickluna's dry dock was at the lower left, and mill sites of T.R. Merritt, James Norris and others were across the water at the right. The Glenridge Bridge was demolished in 1955, and its successor has recently been supplanted.
– J.N. Jackson; NAC: PA-72608

11.15 Homer, ca. 1930: (Opposite) Not all of the effect of the canals on the landscape is immediately obvious, and not all engineers have been oblivious to the impact of new canal structures. For example, Bridge 4 at Homer (the only double-leaf rolling lift on the Welland Canal) has the counterweights placed under the road surface to improve the visual effect. The soaring curve of the Garden City Skyway has arched over this area since 1963, and the Homer Bridge has recently undergone repairs.
– St. Lawrence Seaway Authority: Western Region

11.13 St. Catharines, ca. 1920: In the aerial view of the junction of Ontario and St. Paul streets, the Glenridge Bridge and the Second Canal are prominent. The Canada Hair Cloth factory (centre background, located between the hydraulic raceways), was established in 1888 and was but one of many industries located along the canal [see also Chap. 3, Fig. 3.16]. This area has undergone considerable change: the junction area has experienced several modifications, and Highway 406 has gobbled up much of the valley. It should be noted that the modern highway follows the old canal route. While Dick's Creek, which became the First and Second canals, has disappeared underground at this point, the transportation corridor itself remains a dominant feature of the urban landscape, though the Second Welland Canal long ago ceased to function. – Burtniak Collection

11.14 Port Weller, ca. 1930: The sleek reinforced-concrete structures of the Fourth Canal were influenced by the contemporary "Art Deco" fascination with geometric shapes: note the "streamlined" decorative cornice. Only the upper part of the intake and valve house at Lock 1 is visible during the shipping season. Even the usually submerged structures shown here (the smoothly curved supports and grid-like ducts) are esthetically pleasing as were the elegant stone structures of the Second and Third canals — harmonious additions to the landscape.
– P.J. Cowan, *The Welland Ship Canal Between Lake Ontario and*

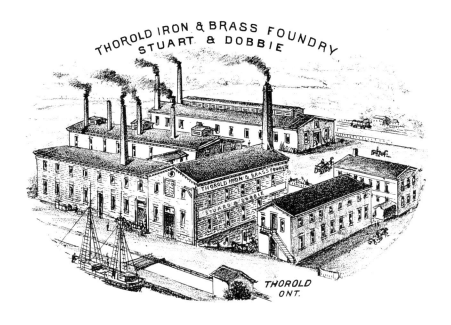

11.16-18 Thorold, ca. 1890-1989: In 1865 Archibald Dobbie established his Thorold Foundry on Front Street. Flourishing in 1890 [Fig. 11.16], the business continued after 1902 under the sole management of John Stuart, providing equipment for many Peninsula industries. By the late 1970s [Fig. 11.17], however, the once dominant feature of this part of Thorold had fallen on hard times, and today [Fig. 11.18] the pre-eminence of Thorold's first foundry is but a memory. – St. Catharines Centennial Library, Photo Album 13, no. 26; R. R. Taylor

11.19 Niagara Peninsula, 1980s: This satellite view shows clearly how the Welland Canal (centre) has affected the landscape of the Niagara Peninsula by creating an artificial waterway to the west of the Niagara River (bottom right). In the first place, the water flow from Lake Erie to Lake Ontario produces noticeable silting at both Port Dalhousie and Port Weller. Secondly, and of profound significance for the economy, the original settlements along the River (with the exception of the crossing points to the United States) have been far outstripped by the newer communities which developed along the Canal.
– WorldSat International, Inc., Mississauga, Ontario

139

Chapter 12: The Heritage Canals

12.1 Welland, 1989: A number of new initiatives are helping to increase the awareness and appreciation of the Niagara Peninsula's rich canal heritage. In Welland, for example, the second Festival of the Arts again featured mural artists who portrayed scenes of the city's past on the outer walls of many local buildings. This program, which began in 1988, now comprises 31 colourful murals, of which ten concern the Welland Canal, past and present. A Port Dalhousie restauranteur recently commissioned Grantham High School students to decorate an outside wall of his establishment with twelve murals depicting historic scenes, the Royal Canadian Henley Regatta and Lakeside Park. – R.R. Taylor

12.2 St. Catharines, 1990 (above right): The new St. Catharines Museum at Lock 3 is part of a complex which includes the viewing stand and tourist facilities. The 600,000 annual tourists to Niagara's second most popular attraction have easy access from the viewing platform to the Museum, which is greatly expanded. The largest of the Museum's five galleries focuses on the history of the Welland Canal. Exhibitions include an exact-to-scale working model of Lock 3, which helps to explain the working of a lock. – St. Catharines Museum, 90-330

12.3 St. Catharines, 1989 (opposite left): In the summer of 1987 Lock 24 of the First Canal was partially excavated, providing historians and canal buffs with much valuable information. In order to preserve the remains of the 1820s wooden lock for a possible later study, it had to be re-buried. However, part of the structure was outlined and a commemorative plaque was erected to mark the site of the original Welland Canal as it climbed the Niagara Escarpment. – Welland Canals Society

$310-million vision for canals

DIG INTO

GEN.L PLANS FOR TIMBER LOCKS

HISTORY

with the Welland Canals Preservation Association

By CAROL ALAIMO
Standard Staff

At least $60 million tax dollars and another $250 million in private investment will be needed to transform Niagara's Welland Canal system into a "first-class tourist attraction," a consultant's study says.

A separate agency, similar to the Niagara Parks Commission, may eventually be needed to oversee development of the plan being hailed as the start of a "new vision for Niagara."

"This is a great challenge, unprecedented in scope in Niagara," regional planning chairman Eric Bergenstein said yesterday as the Region unveiled its Welland Canals Development Guide.

The report calls for creation of a string of tourist facilities, historic sites and attractions stretching from Port Colborne to Port Dalhousie.

The ambitious plan would take more than a decade to become reality, but only if enough money can be found in federal or provincial coffers to foot a large chunk of the bill.

Niagara Region would be expected to fork out $3 to $4 million over the first three years of the development, with some local municipalities to contribute various amounts.

The $200,000 study, paid for by the Region, area municipalities and the federal and provincial governments, estimates the improved canals system could lure as many as two million tourists a year.

That would pump an extra $87 million a year into Niagara's economy and create about 340 jobs, the report said.

Among the major proposals:
● A $14.4-million Transportation Technology Centre near the Flight Locks in Thorold. The facility would include a theatre and "hands-on" exhibits similar to those at the Ontario Science Centre.
● A $36-million transformation for downtown St. Catharines south of St. Paul Street near the Canada Hair Cloth building, with a waterfront to be re-created by refilling the old second canal basin (now the lower level parking lot). The area would be a backdrop for new shops, gardens and displays on local agriculture and industry.
● A $4.6-million facelift for Mountain Locks Park in Thorold.
● About $1.5 million for Lakeside Park in Port Dalhousie, including the addition of "heritage rides and activities" to cash in on the area's nostalgia.
● A $3-million garden in Welland and a $500,000 ship display in Port Colborne.
● Creation of an interim agency to oversee the canal corridor development for three years at an administrative cost of $350,000 a year.

Supporters of the plans say they believe initial public investment will spur private enterprise to join in building many of the proposed facilities.

The report urges development of parks, trails, historic sites, marinas, campgrounds and attractions at numerous sites along the canal system.

It also calls for an eventual $11-million recreational waterway from Port Dalhousie to Port Colborne, connecting to the Niagara River via the Chippawa River at Port Robinson.

A permanent agency may eventually be needed to oversee the projects. The report suggests creation of a Welland Canals Conservation and Development Commission which would operate at an estimated annual cost of $1.5 to $2.7 million.

Al Barnes, president of the Welland Canals Society, praised the study and urged the Region's planning committee to act so it doesn't end up "resting on a shelf somewhere."

Regional planners will recommend at the next regional council meeting that a co-ordinator be appointed to start beating the bushes for funding and mapping out the future of the canal plan.

The development guide will be sent to area municipalities and agencies for comment and be the subject of a public meeting next month.

Canals Drive

12.4 Canals Drive, 1989: A number of organizations, among them the Welland Canals Foundation, and the Welland Canals Preservation Association, continue to promote the waterways in both their operational and historic capacities. In an effort to make our canal heritage more accessible, the Welland Canals Society inspired the erection of a number of distinctive blue and white signs to direct motorists along a two-hour tour from Port Dalhousie to Port Colborne. For the first time there is a recognizable route which takes tourists along the canal, emphasizing it as a major feature of the Niagara Peninsula (a colourful brochure, with map, is available). Incorporated in 1986, the Society represented over a dozen heritage, business, tourism and government agencies in the Niagara Peninsula. In 1987 the Society produced the *The Welland Canals Corridor Development Guide* to serve as a comprehensive strategy for heritage-based economic development in the area. – R.M. Styran

12.5 St. Catharines, 1988: A number of recommendations included in *The Welland Canals Corridor Development Guide* are being pursued by both private organizations and government bodies. While no specific details are available, plans for development at several sites along the canal corridor are being considered. Also in 1988 the City of Thorold purchased property in Fort Robinson (now a part of Thorold) containing locks from both the First and Second canals, thus rescuing from oblivion a significant piece of canal history.
– *The Standard*, St. Catharines, Ontario, 8 September 1988

12.6 Welland, 1972: Various communities along the canal route continue to publicize their canal heritage. Welland's murals have already been mentioned [Fig. 12.1]. Here the Main Street lift bridge is seen floodlit to mark the last transit of a ship (the GEORGIAN BAY) before the By-Pass was opened in 1973. Since then the bridge has been permanently in the down position and is now floodlit during the summer months. – WHM

Activities such as those described above, which help to stimulate interest in the canal heritage, will, it is hoped, prevent future loss of attractive and/or significant canal-related structures, such as has taken place in recent years. The losses we show in the St. Catharines area are representative of others elsewhere in the Peninsula.

12.7 St. Catharines, 1914: The Red Mill, at Lock 4 of the Second Canal, was a landmark from its construction in 1882 until its demolition in 1965. It was on the site of an earlier mill, built about 1828 at the foot of Geneva Street by Oliver Phelps, one of the contractors for the First Canal. It had a long history as a mill before being taken over by Packard Electric in 1895; later it was home to various businesses [see also Chap. 3, Fig. 3.1 and Chap. 4, Fig. 4.1]. – SCHM: N-7337

12.8 Third Canal, ca. 1910: Which one of the many charming — and functional — shanties constructed for Third Canal locktenders this might be, we do not know [see Chap. 10, Fig. 10.4]. The lone survivor, long neglected at Port Dalhousie, has recently been completely restored by the students of Lakeport Secondary School, using as much of the original materials as possible. Removed to the school for the work, it has been replaced on its original site, on a new concrete foundation. It has been more fortunate than the Shickluna houses [see Introduction, Int. 20]: in 1977 a St. Catharines alderman said that saving these small wooden houses built on the Second Canal by Louis Shickluna for his workmen "would rank 169th out of 169 items on our priority list."
– St. Catharines Centennial Library; *The Standard*, St. Catharines, Ontario, 19 July, 1977

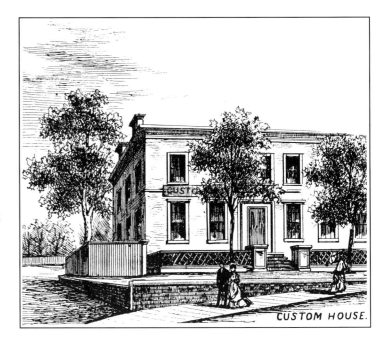

12.9 St. Catharines, 1876: This attractive stone-trimmed brick structure was built in 1855 by the Board of Public Works as offices for the Welland Canal and served as canal headquarters until 1965, when the St. Lawrence Seaway moved to more spacious offices on the Fourth Canal. As indicated here, for a time it also accommodated the Customs House. It barely survived until its centennial: demolition came in 1970.
– SCHM: N-4080

12.10 Merritton, 1980: Housing for locktenders could once be seen at lock sites of the Second Canal [see, for example, Chap. 11, Fig. 11.4]. Only a few years ago there were three of these double stone houses beside Neptune's Staircase on Bradley Street. This one was demolished in 1988 to make way for yet another plaza. – R.R. Taylor

12.11 Port Dalhousie, 1989: The educational and recreational projects being undertaken to enhance awareness of Niagara's canal heritage are potentially profitable. So too is a commercial venture at Lock 1 of the Third Canal in Port Dalhousie, where a new hydro-electric generating station opened in 1989. While this return to the original use of falling water to generate power for industrial use may not be practical elsewhere along the historic canals, it is a welcome reminder of the role of the Welland Canals in the development of the Niagara Peninsula. Unfortunately the pattern of water currents was changed when the generating station went into operation, causing difficulties for pleasure craft. Part of the lock breast wall had to be demolished in an attempt to correct the problem. The Welland Canals Society was able to arrange for an archaeologist to assist in the demolition so that valuable information concerning the original construction of the lock will be obtained. – R.R. Taylor

Even the most ardent supporters of William Hamilton Merritt's "crazy crotchet" could not have foreseen the extent to which the Welland Canals have influenced the Peninsula, but surely they — and Merritt himself — would be pleased with the scope of development triggered by their efforts. "Mr. Merritt's Ditch" was intended at first merely to provide a more reliable source of water power for his mills on the Twelve Mile Creek and to carry small barges or bateaux between lakes Ontario and Erie. The original plan was soon amended to allow the waterway to accommodate larger vessels, even the steamboats which were becoming common. The railway age did not destroy the usefulness of the Welland, nor have later developments in transportation history. The vision of Merritt and his associates, of increasing prosperity in the Niagara Peninsula, has been fulfilled.

GLOSSARY

(With acknowledgement to Arden Phair, Curator, St. Catharines Historical Museum, John Ter Horst, St. Lawrence Seaway Authority, Western Region and E.B. "Skip" Gillam)

ABUTMENT: a solid pier of stone or timber designed to support bridges or other similar works.

AFT: toward the stern of a ship.

APPROACH WALLS: walls of considerable length leading up to locks, to protect locks and gates by allowing vessels to pull up alongside well ahead of the lock itself.

APRON: the platform or sill at the entrance to a lock.

AQUEDUCTS: structures built to carry the first three Canals over the Welland River. The first one, originally built for the Feeder Canal, was of timber; the others were of stone.

ASTERN: movement of a ship's engines to reverse direction.

BALANCE BEAM: see GATE LEVER

BALLAST: heavy substances used to increase the stability of a ship or to submerge the propeller.

BANK: see BERM

BASCULE BRIDGE: a type of drawbridge used on the present canal, with a counterpoise weight which rises or falls as the bridge is raised or lowered.

BASIN: the enclosed part of a dock where ships are moored to load, discharge, or be repaired.

BEAM: one of the heavy pieces of timber or iron set across a vessel to support the decks and to stay the sides; hence the greatest width of a vessel.

BERM: an earthen embankment forming the side of a canal channel.

BOLLARD: a low, thick post, usually of steel and/or concrete, set in the berm along a canal, to which mooring lines are secured when docking a vessel. Also used for controlling a rope or cable that is running.

BOOM (of a lock): see GATE FENDER

BOOM (of a ship): a large crane-like apparatus extending over the deck hatches of a vessel. Used in self-unloading a ship's cargo see SELF-UNLOADER.

BOW: forepart of a ship.

BREAKWATER: a structure for protecting a harbour from the force of waves.

BREAST WALL: a section of wall leading up to the lock gates.

BRIDGE (of a ship): the elevated superstructure with from which the ship is steered and navigated.

BRIDGES: The present canal (Welland Ship Canal) uses the following types of bridges: vertical lift (with a 36.5 m or 120 ft clearance), drawbridges (single span), and double-leaf rolling-lift bascule bridges (jack-knife).

BRUCE TRAIL: a 696.7 km (432 mile) hiking path following the Niagara Escarpment from Queenston in the east to Tobermory on Georgian Bay in the west. It can be found in the vicinity of all the Canals.

BULBOUS BOW: the underwater snub-nosed shape of a ship's bow, designed to break the impact of waves.

BULK CARRIER: a ship designed to transport bulk cargoes such as ore, grain, coal or chemicals. Cranage or other dock-side assistance is required to load or unload the cargo.

BULKHEADS: vertical partitions forming compartments in a ship.

BUOY: a floating object used to mark a navigable channel.

BY-PASS: channel of the present Canal completed in 1973 to by-pass the winding route through the city of Welland. The 13.4 km (8.3 mile) length is straighter, wider and deeper than the former channel and cost $188.3 million.

CANAL: an artificial watercourse used to carry ships; can include improved channels of natural waterways.

CANALLER: a ship designed to fit snugly in the locks of the Second and Third canals to maximize carrying capacity. It had a box-like hull with bow and stern rising almost vertically out of the water. Modern canallers are still built for a similar purpose, with a maximum length of 222.5 m (730 ft), and are easily distinguishable from the "salties."

CAPSTAN: an apparatus for hoisting anchors by winding a cable around a rotating drum.

CHAMBER: see LOCK

COFFERDAM: a temporary enclosing dam built in the water and pumped dry to protect labourers while

performing some work, such as the construction of pier foundations, is in progress.

COPING: a top course of a lock wall, of stone or steel, sloping to shed water.

CRIB: a wooden or concrete frame upon which to build a wall or pier.

CULVERT: any artificial covered channel for the passage of water under a road, canal, berm, etc.

CULVERTS (for locks): filling and emptying culverts for the Welland Ship Canal (Fourth Canal) locks are 4.25 m (14 ft) wide and 5.03 m (16 ft 6 in) high. One on each side of the lock leads to 25 branch openings along the base on each side of the lock, running the full length of the chamber.

CURTAIN WALL: that portion of a wall which connects two advancing or more lofty portions.

DAM: a barrier of earth, wood, stone or concrete for holding back or confining water.

DECKS: platforms within a ship's hull extending from side to side and from bow to stern.

DEEP CUT: the high section of the Welland Canal between Allanburg and Port Robinson. Referred to as the "Deep Cut" because additional excavation was necessary to bring it to the water level of the Welland River. A tunnel 5.05 m (16 ft 8 in) wide and 5.03 m (16 ft 6 in) high was originally proposed for the channel of the First Canal in this section. Eventually a "deep cut" of 21.33 m (70 ft) was dug. Unfortunately, when the work on the First Canal was nearly finished, earth slides occurred, setting back completion by nearly a year. A decision was then made to supply the Canal with water from the Grand River rather than the Welland River, as originally planned.

DRAFT/DRAUGHT: the depth of water that a ship needs for floating, or the depth to which a vessel is immersed when bearing a given load.

DRY DOCK: a basin or chamber with watertight gates for controlling the water level, used in the construction or repair of boats.

EMBANKMENT: see BERM

FEEDER CANAL: the Deep Cut (between Allanburg and Port Robinson) could not be excavated down to a level where its water supply would be derived from the Welland River. Instead, the Grand River was dammed, raising its water level, and a Feeder Canal was dug from the Grand River at Dunnville to the Deep Cut, a distance of 43.45 km (27 miles). It opened on September 28, 1829, with a depth of only 1.5 m (5 ft) and a bottom width of 6 m (20 ft).

FENDER: any resilient material or bundle of material used to protect the side of a ship or other floating structure from damage; usually suspended on a rope.

FITTING OUT: completing the construction of a ship, usually after launching.

FLIGHT LOCKS: on the present canal, a series of three locks which are twinned in flight to allow the simultaneous upward and downward movement of ships. Their close proximity to one another gives them the appearance of a giant staircase. Locks 4, 5 and 6 complete a lift of 42.6 m (139.5 ft).

FLOATING TOWPATH: a floating towpath on a double row of wooden pilings was built for the first three Canals in order to cross large bodies of water. Tow animals walked on these paths when pulling vessels through the canal. In certain stretches (as at Port Dalhousie), a section of the path could be removed to allow the passage of vessels.

FLUME: a narrow channel to carry water from a canal, river or raceway to drive a mill wheel.

FOGHORN: sounding horn used as a warning during fog.

FORE AND AFT: lengthwise direction of a ship.

GALLERY: see MOORING PASSAGES

GATES: moveable barriers swinging on hinges, for closing a passage into a lock. The lower gates of the present Canal each weigh 453.75 tonnes (500 tons) and measure 25 m (82 ft) in height.

GATE CHAMBER: the recess in a wall into which a gate may swing so that it is flush with the wall when fully opened.

GATE FENDER: a steel wire rope with a diameter of 8.8 cm (3 1/2 in) on a 269 m (82 ft) steel boom extending horizontally across the lock chamber of the present Canal at coping level to prevent the accidental ramming of a gate by a ship.

GATE LEVER: a man-powered device likely employed on the First and Second canals to open and close the lock gates. Large beams acting as levers extended out from the top of each gate and were pushed or pulled to operate the gate.

GATE YARD: a canal maintenance yard for building or repairing gates or other lock mechanisms.

GONGOOZLER: an idle and inquisitive person who stands staring for long periods at anything uncommon, especially at canal boats and canal people.

GOWAN SAFETY DEVICE: a casting attached to the end of the gates which prevents them from becoming detached from the mitre (Third and Fourth canals).

GRAND RIVER NAVIGATION COMPANY: a shipping company formed in 1831 to transport goods down the Grand River. Like the Welland Canal Company, it also obtained hydraulic privileges, and was promoted by William Hamilton Merritt as a means of opening up markets and providing better transportation for communities as far up the Grand River as Galt. Dissolved in 1861 due to foreclosure.

GRAVING DOCK: a dry dock for examining, cleaning and repairing a ship's hull.

GRAVITY: the Welland Canal locks are emptied and filled by the force of gravity when valves are opened in the lock walls.

GUARD GATE: placed above Thorold at the beginning of the Summit Level on the Second, Third and Fourth canals as a safety gate to prevent flooding in the system below it.

GUARD LOCK: the present Guard Lock at Port Colborne is one of the largest locks in the world at 420.6 m (1,380 ft). It compensates for the varying water level of Lake Erie and the regulated level of the Canal system.

HARBOUR: a sheltered area for a ship, either natural or man-made.

HARBOUR MASTER: official in charge of the harbour and regulations governing it.

HATCH: an opening in the deck of a ship affording passage to the cargo hold.

HAWSER: a rope of steel or fibre for mooring, hauling or towing.

HEAD GATES: the upstream gates of a canal lock.

HEADRACE: see FLUME

HEELPATH: the side of the canal bank opposite the towpath.

HEEL POST: the post in a lock wall to which the gate is hinged.

HELM: the mechanism by which the rudder of a ship is controlled.

HENLEY COURSE: permanent home of rowing for the Canadian Association of Amateur Oarsmen since 1903. Uses the area of Martindale Pond in St. Catharines following the First and Second canal routes.

HOLDS: compartments in a ship for "holding" the cargo.

HOLLOW QUOIN: the vertical recess in the lock wall into which the end of the gate leaf fits.

HYDRAULIC COMPANY: in March 1831 the Welland Canal Company sold its landed property, together with the rights to sell or lease the surplus water, to a group of men headed by John B. Yates. This was a very attractive arrangement, in light of the potential water privileges which were available. Repurchased by the Welland Canal Company in June 1834.

HYDRAULIC RACE: a man-made channel that carried water at a higher level than the canal. In falling to a lower level, the water provided power for mill wheels. The system paralleled the first two Welland Canals from Thorold to St. Catharines, and was dewatered and abandoned after 1929.

INTAKE: the point at which water is taken into a pipe or channel.

INVERTED SIPHON CULVERT: six 6.7 m (22 ft) siphon tubes carry the Welland River under the Ship Canal at Welland.

JIB: a derrick or crane with a pulley at the upper end for lifting heavy objects.

JUNCTION LOCK: a lock which joined the Feeder Canal to the first three main canals at Welland; also, a lock joining the Third Canal and the Welland River.

KEEL: basic part of a ship's construction and the first to be laid down when building a new ship. Runs the full length of the ship's bottom.

KNOT: nautical unit of speed equivalent to about 1.85 km (1.15 statute miles or 6,080 ft).

LAKER: a ship intended to serve primarily in the Great Lakes. Many conform to the dimensions of the Welland Canal locks (max. length 222.5 m or 730 ft); others may now have a length of up to 305 m (1,000 ft.) Most have a high bow and low stern.

LEAF: a division or part of a pair of lock gates, or bridge spans. See BRIDGES and GATES

LEVEL: see REACH

LIFT: the distance which a ship is raised from one reach to another by means of a lock.

LIGHTERING: unloading all or part of a ship's cargo to reduce its draft.

LIGHTHOUSE: a tower containing a light for guiding navigators by night or during inclement weather; erected at the entrance of a port or at a point of danger.

LINESMEN: the people who secure the mooring lines of ships.

LOCK: an enclosure with gates at each end used in raising or lowering ships as they pass from one water level to another.

LOCKMASTER/LOCKTENDER: canal worker responsible for locking boats into the canal, opening and closing gates, recording ship passages.

LOCKTENDERS' HOUSES: semi-detached stone housing provided by the government for Second Canal locktenders.

LOCKTENDERS' SHANTIES: small frame buildings erected at the Third Canal locks to serve as offices and sleeping quarters for the locktenders.

LOWER GATES: the downstream gates of a lock.

MAST: a long vertical pole of wood or tubular steel, originally for carrying sails, but now used for navigational equipment.

MERRITT'S DITCH: early name given to the First Welland Canal, a small winding canal which followed creeks wherever possible.

MERRITT'S FOLLY: a derogatory name given to the First Canal by those who scoffed at the practicality of Merritt's waterway.

MILLRACE: see HYDRAULIC RACE

MITRE: the angle of closed lock gates, pointing towards the upper water level.

MITRE SILL: wood or stone structure along the bottom ledge at each end of a lock, shaped in a 'V', pointing upstream, upon which the lock gates close.

MONOLITH: strucure of the present Canal formed of one large mass of concrete.

MOORING PASSAGES: openings in the walls of the locks of the present Canal, 8.8 m (29 ft) below the coping level. These passages were intended to be used in a two-stage mooring of vessels, but this plan was later abandoned as unnecessary and the openings are now being filled in.

MOORING POST: see BOLLARD

MOUNTAIN LOCKS: the series of locks of the Second and Third canals which were "shelved" into the side of the Niagara Escarpment in their ascent of the mountain.

NAUTICAL MILE: see KNOT

NEPTUNE'S STAIRCASE: name given to the tightly grouped series of Mountain Locks (#16-21) of the Second Welland Canal in reference to Neptune, the Roman sea god, and to their "staircase" appearance.

This phrase, used during a visit by Lord Elgin and a party of government officials in 1850, originated during construction of the Caledonian Canal in Scotland.

PADDLE: see SLUICE

PASSING BASIN: wide point in the canal channel to enable approaching vessels to pass one another. The larger or upbound vessel was allowed first passage.

PENSTOCK: a huge pipe or conduit which carries water from the head gates of a weir pond, raceway, hydraulic race or canal reach to power a water wheel or turbine.

PIER: a man-made structure extending from land into a water body, acting as a breakwater to provide a safe point of entry for vessels approaching the canal system. Also, any solid support for the ends of adjacent spans in a bridge.

PILING: heavy pointed timbers or steel beams forced into the earth to form a foundation for a wharf, building, or the like.

PILOTS: qualified men licensed by the local Seaway Authority to board ships (usually foreign ocean-going vessels) at each end of the Canal, to direct the master of the ship in going through the unfamiliar system. Their knowledge of local conditions is used to help the master, who is not relieved from his duties or overall responsibiity. Canadian and U.S. flag vessels which normally trade within the Great Lakes may operate without a pilot.

PINTLE: the pin upon which a lock gate swings.

PIVOT POST: see PINTLE

POND: a basin of water built up behind a regulating weir.

PONTOON GATE LIFTER: a vessel used during construction of the present Canal to place a gate leaf in position in a lock. It was specially designed to manoeuvre in confined spaces and had a blunt end with lifting frame to handle gates. Watertight bulkheads at each end of the vessel were filled or emptied to compensate for the 453.72 tonne (500 ton) weight of the gate.

PORTAGE: the act of transporting boats and cargo overland from one navigable body of water to another.

POWERHOUSE: generating station at the foot of Lock 4 which develops 10,000 h.p. of hydro-electricity from the 56.7 m (186 ft) fall of water carried

by a penstock from the head of Lock 7. This power is used in the operation of the present Canal.

PRISM: the cross section of the canal channel; also used to refer to the channel itself.

QUAY: a landing stage built alongside a basin or harbour for loading and unloading cargo.

RACEWAY: see HYDRAULIC RACEWAY.

REACH: the level expanse of water between locks.

REGULATING WEIR: a dam-like structure which maintains a correct level of water in the pond and/or reach above a lock.

ST. LAWRENCE SEAWAY AUTHORITY: a Canadian Crown Corporation established by Parliament in 1954 to construct, maintain and operate a deep waterway between Montreal and Lake Erie. Assumed operation of the Welland Ship Canal in 1959.

ST. LAWRENCE SEAWAY DEVELOPMENT CORPORATION: formed in 1954 by the Wiley-Dondero Act. The U.S. Congress authorized the St. Lawrence Development Corporation to join the St. Lawrence Seaway Authority in constructing navigational facilities in United States territory.

SAFETY WEIR: a weir used in conjunction with a Guard Gate to regulate the water in the event of an accident.

SALTY: an ocean-going vessel, as opposed to a "laker."

SCHOONER: a small sea-going sailing vessel with two, three or four masts carrying one or more topsails.

SCREW: a propeller-driven ship, as opposed to a sailing vessel.

SELF-UNLOADER: a ship designed to transport bulk cargoes, capable of unloading its cargo directly onto a bare dock or into a hopper, without dock-side assistance. The unloading of a laker requires only about 4 1/2 hours, as opposed to 18-24 hours for a bulk carrier.

SHIPS' CHANDLER: a dealer supplying groceries, provisions, hardware and ship's stores to vessels passing through the canal.

SILL: see MITRE SILL

SLAB: protection on side banks of the canal.

SLUICE (paddle): a frame of timber or metal within a lock gate for the purpose of regulating the flow of water.

SNUBBING POST: see BOLLARD.

SPILLWAY: a passageway in or around a dam to release the water in a reservoir.

SPOIL: excavated material.

STONEY VALVE: a type of vertical valve on rollers, used in various weirs of the present Canal.

STOP LOGS (or GATES): gates placed at intervals along a waterway and usually kept open, but which can be shut when required, to isolate a stretch of water containing a breach of the walls, or when repairs of any kind are necessary. Originally a stack of logs kept beside a canal to form a gate.

SUMMIT LEVEL: the highest level in a canal system from which no further lifts are necessary. Occurs just south of Thorold between Allanburg and Port Robinson.

SURGE TANK: a storage reservoir placed at the downstream end of a feeder pipe at Lock 4 to control variations in water pressure.

TAILRACE: a channel for returning "used" water from industry to the canal.

TAINTOR GATE VALVES: valves installed at the intakes and discharges of canal locks and also in certain weirs. These valves pivot in an arc of a circle to regulate the flow of water.

TELEMETRY: the determination of distances by use of a telemeter.

TIE-UP WALL: see APPROACH WALL.

TONNAGE (deadweight): tonnage of a ship representing the weight of cargo, bunker fuel, water, passengers and crew.

TONNAGE (displacement): actual weight of a ship.

TONNAGE (gross): the carrying capacity of a ship expressed in tons of 100 cubic feet of enclosed space.

TOP HAT: at the commencement of the shipping season, a top hat is awarded to the captain of the first upbound vessel to enter the Canal from an outside port.

TOWBOYS: men who led teams of horses or mules pulling vessels along the First, Second and Third canals. With the introduction of self-propelled ships and tugs, they were no longer needed.

TOWPATH: path along one bank of the canal where teams of horses or mules walked while pulling ships.

TRAFFIC CONTROL CENTRE: the St. Lawrence Seaway Authority office uses computers and closed-circuit television to improve the scheduling of vessels and to reduce the overall transit time of ships through the present Canal.

TRANSIT: action of passing through the canal.

TUGBOAT: a small, powerful boat used to push or pull large ships.

TURNING BASIN: an area of water wide enough to allow vessels to turn around and reverse their direction. For example, a turning basin was used by excursion boats travelling up the Second Canal as far as Lock 2 at Wellandvale in St. Catharines.

TWIN LOCKS: see FLIGHT LOCKS.

UNWATERING: freeing the canal from water. Necessary when inspecting or repairing lock mechanisms.

VALVE: any contrivance that opens and closes an aperture. See, for example, PADDLE or TAINTOR.

WASTE WEIR: a dam-like structure along the canal berm with openings for controlling the water level. Water not required for the operation of the locks is diverted to the weir and pond system which parallels the canal locks.

WELLAND CANAL COMPANY: incorporated on January 19, 1824, with George Keefer, President, and William Hamilton Merritt, Financial Agent. Private company with capital set at £40,000 to construct an artificial water channel between Lake Ontario and Lake Erie. In 1843 an act was passed by Parliament to compensate the investors in the company, an action which effectively signified the end of the private company and the canal's takeover by the government of the United Canadas.

WELLAND CANAL LOAN COMPANY (Welland Canal Manufacturing Company): in 1851 several local businessmen petitioned for incorporation of a company to purchase the Welland Canal lands between St. Catharines and Thorold, to encourage the erection of factories and mills. This land was then subdivided and sold for industrial, retail and residential purposes. The town of Merritton was developed by this company, which was dissolved in 1930.

WELLAND CANAL LOCK COMPANY: formed by Oliver Phelps, a major contractor on the First Canal, to carry out construction on the Canal.

WELLAND RAILWAY: a railway developed between 1853 and 1858 in conjunction with the Second Welland Canal. It was designed to take cargo from ships entering the Canal in order to reduce their draft and to enable them to navigate the Canal. It also operated in winter, when the Canal was closed. See also LIGHTERING.

WINCH: a cranking mechanism which opened or closed valves in the lock gates in order to fill or empty the locks of the first three canals.

WINDLASS: a horizontal drum for winding or for hoisting by winding.

WING WALLS: walls which are flared at an angle forming the approach ends of the locks.

INDEX